© Copyright 2024, Prasad Prakash Tupache.

All rights are reserved. No part of this book may be reproduced or transmitted in any form by any means; electronic or mechanical including photography, recording or any information storage or retrieval system; without the prior written consent of its author.

The opinions/content expressed in this book are solely of author and do not represent the opinion /standings/thoughts **Amazon Kindle Direct** Publication. No responsibility or liability is assumed by the publisher for any injury, damage or financial loss sustained to a person or property by the use of any information in this book, personal or otherwise, directly or indirectly. While every effort has been made to ensure reliability and accuracy of the information within, all liability, negligence or otherwise, by any use, misuse or abuse of the operation of any method, strategy, instruction or idea contained in the material herein is the sole responsibility of the reader .Any copyright not held by the publisher are owned by their respective authors. All information in this book is generalized and presented only for informational purpose "as it is "without warranty or guarantee of any kind.

All trademarks and brands referred to in this book are only for illustrative purpose are property of their respective owners and not affiliated with this publication in any way .The trademarks being used without the permission don't authorize their association or sponsorship with book.

ISBN : 9798341220256

Price

Publishing Year: 2024

Published and Printed By

Independently Published Through Amazon Kindle Direct Publication

Office Address: Amazon (India) Brigade Gateway, 8 Th Floor, 26/1, Dr. Raj Kumar Road , Malleshwaram (W) ,
Bangalore – 560055
Phones: +918033273000
E-mail : amznindpr@amazon.com
Website: www. Amazon.in

Printed in India & Various International Amazon Marketplace (Website) Through Print on Demand Technology

WITH BEST COMPLIMENTS !

M/S TUPACHE CONSULTANTS

PROPRIETOR: MR.PRASAD PRAKASH TUPACHE

UAN: MH26D0030607

SURVEY NO 79/20, SHIV RATNA COLONY,

PACHPIR CHAUK, KOKANE NAGAR, KALEWADI, PIMPRI, PUNE-411017

CONTACT: 9970173983

ABOUT THE AUTHOR

Author : Mr. Prasad P. Tupache ,

Address: Survey No 79/20,

Shivratna Colony , Pachpir Chauk,

Kokane Nagar, Kalewadi ,

Pimpri, Pune – 411017

Font Setting :   **Publisher** : Amazon KDP

Mr. Prasad P. Tupache .

Cover Design :

Mr. Prasad P. Tupache .

Photo Courtesy :

Front Page : Julie Mac -Pixabay , Background- Canva.com

Rear Page : Background- Canva.com

Created From : Canva.com

BLESSINGS

Image Courtesy: Tarun Shihora, Pixabay.com

Namaste Friends,

Welcome to the blessing page of our new project – The Scientist -Stories beyond inventions!

The journey of writing on subjects of interest is always a fascinating task ! You know something and you share it with others ! Others read it and either accept it or suggest some changes based on their knowledge of that subject ! This knowledge exchange adds up to authors knowledge as communicator and this exchange keep happening as you keep writing again and again !

A point comes into life of any author where they always feel blessed to receive so much power of imagination as they go on writing seamlessly ! The power of imagination when get associated with feeling of sharing important facts that build knowledge and awareness always prove to be surviving factor in the times of challenges and turbulence !

During this journey up to 14 projects , we sincerely thank 'Bappa' for his constant inspiration to gain knowledge and stay humble till you do not complete your project ! Acquiring knowledge is like wearing jewelry for every scholar ! It simply beautifies your intellect and make your society presence lively ! People can easily say , if this person is present in this group, then the group will get good updates about dynamic changes happening in and around ! This type of confidence is received only after continuous study and for this study 'Bappa' always show the way where light is falling sharp and clear !

As this project is about 'The Scientist ', we would like to remember few lines about knowledge and science ! Everywhere around the world knowledge is available ! Earlier it was noted by few science enthusiast, they carried out experimentation on these natural observations, they prepared apparatus with which they can try many trials, nothing was present there as readymade resources, it is designed and developed by these science enthusiasts !

So, what was the theme behind this development happened over many years ? It's the tenacity of scientists to serve this society and humanity ! Few people are blessed with brilliant intelligent quotient and hence it is destined for them to use that skill for resolving major live society issues and problems by inventing something which is not present there ! This power to invent is nothing but blessing of knowledge and overall look out towards society needs !

So, on concluding part of the blessing page of ' **The scientist – Stories beyond inventions'**, lets pray for total freedom, happiness and liveliness in the society by utilizing strength of science and its different useful inventions !

DEDICATION

Image Courtesy: Susan , Pixabay.com

This book is sincerely dedicated to all Indian and Global Scientists who always believed in their intellectual capability and made this world easier to live !

This book is also dedicated to all those numerous failed experiments & efforts which were carried out with all possible knowledge application but still missed that spark which ignite the knowledge fire in the belly of researchers ! That spark which specifically gives birth to an invention !

THANK YOU

Image Courtesy : Graphisty , Pixabay.com

Dear All,

We sincerely thanks to all image contributors & creators on **pixaby ,unsplash and other stock photo websites** ! Because of these images , content become easy to relate !

The images used in the project are stock images and as per the website guidelines of usage of these images , the name of the creator is mentioned after inserting particular image !

Every effort is done from our side to justify the image usage and its subject matter to keep the content easily relatable for everyone !

Thank You , once again !

INSPIRATIONAL

Image Courtesy : Amarin , Pixabay.com

INSPIRATIONAL

Image Courtesy : Open clip art , Pixabay.com

PREFACE

Namaste Friends,

It gives me immense pleasure to write the preface of our 15th project – The Scientist : Stories beyond Inventions !

Scientists are known as clever people with golden heart ! Why ? Because ,they study for most part of their life by sacrificing many materialistic pleasures and devote their productive time for inventions which changes the lifestyle of human being and also improves longevity of huge population ! Because of this invaluable contribution to society , scientists are considered as people with Golden Heart !

If you decide to just glance through the impressive lifetime of a scientist's life , you will be amazed to see their dedication of study and research . For our sake of clarity , let's revive the typical life phases of a scientist's life !

Early Spark :

Many parents are fortunate as they are parents of sharp and intelligent kids ! It might be inheritance or nature's wonder ,but some kids are born with exceptional intelligent quotient ! Since childhood , their language skills , verbal communication skills, reasoning skills, computational skills, memorizing skills, conditional adaptation skills , light hearted human nature , funful attitude towards simple things in life , exceptional questioning brilliance and unmatchable curiosity is considered as early signs of future scientist in such kids !

On the other hand , there are several parents who are highly educated and they have great faith in education ! The kids of such parents are not as intelligent as the 'born intelligent kids ! These kids get a supportive surrounding around them in such a way that by the time they attain age of 20-21 , they get enough competence to enter into the field of scientific research and advancement for future innovation! Their further progress takes place with great efforts taken during particular studies !

Beside these two types of research students , there is third type of researchers who are known as researchers by passion ! These kids are quite average in their school time ! In college life also they perform to satisfactory level and become eligible to attain a high paying job ! The twist in their life takes place when they enter a highly competitive professional environment !

In this highly competitive environment , they can see ,how the students less than their age are entering the field with research qualifications and how their career is taking a major leap ! They can see , how the new students are getting big roles and how their contribution is supporting the organization in big way !

This fact makes them highly upset in positive way and they decide to take a break in their professional life after nearly a decade long service and they pursue master's education for some time which is followed by a full-fledged research programme or fellowship in reputed international research institute ! This is the third category of research students who enters after practical field experience !

The next variety of research student consist of pure academic students ! Since childhood , they are interested in learning and acquiring knowledge of different fields ! They keep learning and their family set up also doesn't insist them to have early work experience . They receive full family support to complete their education up to doctorate and in some cases post doctorate level after which they can either become part of their traditional business or they can start their own business !

These kids never leave college atmosphere ! They like to hang around college surrounding and different city libraries ! They are fond readers and they like to collect literature of their field of interest ! They like to interact with their professors and have a good forward going academic relationship with which they receive able guidance and thus improve their research potential day by day !

As they are interested in academic environment , they like to teach undergraduate student and at the same time they pursue their own masters and research education !

After research , they can further carry out post doctorate research to make the things more expansive and impactful ! In this learning and knowledge acquiring process , they become highly qualified research guides for several other students !

Some other research entrants are lads from different work zones ! They have some special learning ability ! They enter into particular research field just because of their newly developed interest in that field ! These people are great contributors to research field !

To explain this point, let's take an example of a heart specialist who has fond interest in Indian classical music ! Since his childhood and may be by age of 7 or 8 , he has started getting musical lessons and improve his vocal singing skills ! His academic progress is also happening along with and at the same time he is also mastering his passion to learn Indian classical music ! At the age of 35/40 , he is well settled in his life and now he decides to carry out research in Indian classical music to find out 'ragas' that heals painful experiences of diseases !

Can the music therapy heal diseases ? Can a doctor prescribe practice of listening different ragas to mobilize the body fluids in positive way ? Can a doctor practice various 'ragas' to balance his professional stress levels ? Can a musical symphony make a patient recover faster from major illnesses ? Just see the chain of possibilities that brings the union of interest in medicine and music for a common student who is a doctors and musical enthusiast also !

Such type of research will have different contribution to society and such kind of researchers are always known as ' difference makers ' in the huge human society !

We can take another example of an IAS or IPS officer who is on deputation to carry out research in 'New ways of urban and rural pollution control !' sponsored by his territory of administration ! Now the current designation of the student is related to public service but the research being carried out is in environmental science ! Then why this officer is recommended by the sponsoring authorities ? The answer of this question lies in previous qualification as well as his previous work experiences ! Before joining the public services , the officer was regional manager of an environmental product producing global brand and he also held master's degree in environmental sciences including study of recent and historical pollutants !

To utilize this professional experience to live public pollution issues and to add advanced knowledge to his kitty , sponsoring authorities has recommended him for an international research programme with institute of repute so that he will be part of global network as regional representative and global knowledge bank member of that specific field !

As the list of various research entrants is going on , new students need to mention here ! These students have done everything in life , they have completed formal degree level education , they have done their job throughout their life , they carried out their family and personal responsibilities and now they have just nothing to do in life ! They are beholder of good health and now they want to explore their vision in the research field of their interest to make their professional qualification bright & brilliant ! This work they want to do till they are living this life ! They can readily spend this time and along with this they can interact with new bright students to have a mutually rewarding learning experience !

The other type of students are proven performer of their field and they do extensive research when a typical role is offered to them ! They literally explore the lifestyle of such roles and they make suitable changes in their expressions to suit that role ! Such field includes students from drama and cinema !

When they have to do a particular role for play or movie , directors may provide some basic material required for role but to understand the character with its fine details , they need to be part of that troop or that group whereby they can adopt required skills , intellect , linguistic brilliance and stuffs like that ! When their roles make wonders in public , these students are felicitated by various universities with honorary doctorates to recognize the way they put themselves in those roles for that project !

Again, when new project is signed , they have to again learn the character , its background , its way of expression and the theme of the project ! So as an actor and as a person ,they are learning many things straight from the present or past society ! If you are doing a modern role , your learnings are part of modern society ! If you are portraying a historical character , then you have to study relevant periodical literature and you have to visit those highlighted places to understand 'what was happened there before 300 or 400 years ago from the local masters and guides ! So, this is also extensive research and only after such type of research , the role played by actor become memorable and true to original expectations that gives both creative and business success to that project and to that actor for long time !

So , one can easily see that research is such a broad concept where you can enter at any point of time if you have previous qualification necessary for this entry ! You can carry out any type of research just to invent something new and useful for society ! You can also carry out research on things which are harmful for society and explain the harmful effect with fact finding so that people will stay away from such happenings ! So ,either by finding a solution to live problem through an invention or by making people alert about harmful effect of their surrounding issue , researchers usually address huge public issues ! Hence ,any progress done in this direction is considered as boon for mankind and hence researchers are highly respected individuals in human societies and professional bodies !

Lifespan :

So, in this preface , we have seen how the early spark in a scientist or researcher is noted by society and rest of the work done is directing this spark to make a controlled stream of fire that simply burns ignorance and enlighten people's lives with gentle and careful approach !

So , how a typical lifespan of researchers is fulfilled with path breaking inventions and its commercial public application is well known to society since past hundred or many centuries , the life of a researcher is life lived for others in true spirit !

Once their research is completed , they have to publish it through science journals and research institutes . The research papers will be read and analyzed over there for its genuineness . The data and formulae derived during research are cross questioned and their scientific proofing is duly interrogated by field experts ! This is a typical testing time for any researcher . He put every reference used during research work to make his stand clear !

The detailed project report written with support of basic theory , assumptions , sample study , proto type making , issue analysis and further high-end technical evaluation is all become part of the project report ! The research projects which are part of human study and directly deals with real time information and interaction with native people and root causes of local painful issues are found out ! Researchers put their all energy to derive a meaningful and impactful solution to this live native issue ! The solution needs to be implemented for sample region in that locality and its effects are analyzed for pre-determined research time !

When there is issue , then there is change to normal ! This change can be considered as deviation to normal or exception to normal. Every researcher has to find out the in-depth mechanism of this abnormality or exception so that they can control the variation , mitigate the risks of that abnormalities , avoid the recurrence of that abnormality or motivate that abnormality into some constructive application ! Yes , researchers have to apply their mind in all possible directions from where they can get the exact required invention !

So, for this logical and scientific study , researchers have to carry out extensive experimentation ! When it is experiment , everyone in the field of science are aware that experiments are done for curiosity of finding something unknown ! So , what is that unknown is the first question that comes to mind of researcher ! Then they choose the body of material or matter where the study is to be concentrated ! When the material is fixed , they choose the process with which they will observe the behavior change in material !

These observations can be effect of heating on material , effect of cooling on material , effect of storing of material under specific atmosphere , effect of handling material in its intermediate phases of transformation ! So , basic observations of any experiment are done to study the effects of experiments !

Every experiment has a typical procedure with which material and matter interact with each other in specific proportion or calculated amount to show the possibility of favorable , unfavorable or neutral reaction ! Researchers has to find out the activity level for various reactants and reaction agents ! They have to study the chemistry and thermodynamics behind every such reaction and convert those observations and findings into mathematical formulations !

Researchers has to choose typical mathematical model with which their observed phenomenon can be suitably formulated !

As every researcher is fairly conversant with required mathematical theories and applications like algebra , geometry , calculus , statistics , logic , series , probability , graphs , they choose the most suitable model to present their finding and their interpersonal mathematical relationship !

In many scientific experimentations , researchers reveal a specific value which is constant throughout various levels and phases of experimentation ! This constant is often named after its first inventor ! The constant play's vital role in determining scientific fundamental relationship between various experimental variables ! Rather ,one can say that because of constant , one can understand the variables effectively and hence manage them in real time applications !

So, when experimental finding is done , same experiment is done with either reducing the sample size or increasing the sample size to find out the range of that effect under available environment ! e.g. if a typical scientific effect is observed at 'x ' value , researchers go far up to 'x

+m ' and ' x-n' to find out the range 'm to n ' under which that effect is positively seen , beyond this range the occurrence of scientific effect does not match with it mathematical connotation !

In many studies , beyond a particular range ,either the first invented effect gets stop or it get converted into some other forms ! This phase transformation is another discovery for researchers ! To specifically explain with the help of example ,one can see different phases observed in steel transformation at different temperatures ! Only after attaining these temperatures till complete melting of steel , researchers could invent presence of ferrite , pearlite , austenite , bainite , martensite , cementite ! The point here is , researchers need to find out the starting and closing range of particular scientific effect so that it can be formulated correctly and with that reference it can be developed for usefulness for mankind !

So , after completion of experiment , a typical inference is derived and that inference is recorded as goal or finding of that experiment ! So , when study of heating of steel is carried out,

One can draw inference that steel melts at 1547 degree Celsius ! The value 1547 degree Celsius gives the accuracy for that experiment and whenever and wherever given grade of steel is heated as per experimental condition , the same truth will be revealed at all experimental sites ! This again proves the fact that temperature is intrinsic property of the material and hence its same at universal level !

To cross question the experimental finding , people may ask -can steel melt at 1400 degree Celsius in hot region and 1648 at cold region ? The experiment proves that steel melts at 1547 degree Celsius only ! Thus, scientific findings are proven universally and then they become universal truth or more specific -scientific laws !

The proven laws are named after their inventors and their name is published with their invention ! This fame received after path breaking inventions resolves issue of that particular field and this invention can become base for further studies carried by other researchers ! So ,this is win-win situation for researchers across the globe !

So, here the role of science journals and publications comes into play ! Because of widescale publicity , the invention reaches masses and scientific discussion starts occurring . This triggers the path of further progress for that field and with valuable contribution of various international scholars and experts , the simplicity of subject goes on rising !

The fundamental purpose of every research is to make things easy for mankind by observing the relationship of natural phenomenon occurring since long long time ! If you know this relationship quite well , we can convert same into application for human use !

In science conference or science congress , many exceptional dignitaries share their insights and experiences with the audience present there ! This intellectual interaction is very much forward looking and it inspire young mind to research with great caution and by availing latest technology accompanying them !

Technology boosts the speed of research . The scientific findings , if they are not confidential , can be shared with peer group and their opinions can be heard ! These opinions further simplify the research and make the research paper broader and more inclusive !

So, one can see , a typical researcher can invent one invention in their whole lifetime or he can invent hundred inventions throughout their lifetime ! It depends upon scientific caliber and field of study in which the researcher is working ! This has profound effect on research results !

Every research has some type of impact on society ! If the history of past research is considered , it is seen that because of research in particular field , development is happened and because of that development, interaction within that region and outside that region is boosted which has resulted into exchange of money , power and facilities !

To understand this through an example , lets understand the recent invention of computers ! When computers become part of official work practices , it is seen that accuracy , speed and sharing of work-related documents is phenomenally improved. When this invention is marketed in outside countries ,many countries noting its usefulness adopted in their working environment . Soon every interested country accepted the digital way of work and thus these inventions revolutionized the earlier way of official working !

Let's take another example of invention of various vaccines ! Vaccines are invented to fight against number of viruses present in the environment. Presence of deadly viruses causes threat to people's life and hence you need a protective mechanism inside your body so that you can build immunity against harmful effects of these viruses !

Researchers of health sciences continuously study the behavior of typical virus and understand the mechanism of its birth , its source place from where it originates , its overall characteristics and symptoms of its infection , how does it get transfer inside human body , which part of the human body are immediately affected and how serious its overall presence for this society ! Based on these practical factors, vaccines are developed with which required antibodies are developed which can deal with harmful effect of viruses !

Based on the intensity of infection , particular time bound doses are prescribed and this is how infection is handled ! The created vaccine is transported to various countries and thus people's life is saved !

So , this is the power of one invention and for this powerful performance ,researchers are awarded with global recognition ! Research is a thing which need hefty investment of financial resources in the start to carry out high end technology assisted experimentation but once that experiment become successful and its pilot model is prepared then the next path completely consists of success from all directions ! Invested money get recovered in first few months of sales of that new invention !

Every invention is life learning for a scientist and it is always a brainchild for him ! That invention gives international recognition to him or her and this fact inspire them to add more value to this society ! When we consider the evolution of scientific concepts life electromagnetism , gravity , mechanical force , computer logic , chemical affinity , radioactivity of material , metallic and non-metallic bonding , we come to know that science is such a vaste field of study present in nearby nature ! Lots of mysteries are there in this nature and it is the curiosity that resolves these mysteries and make them useful for society !

The application of invention is equally important as its protection ! When you invent something new and communicate to world around you , everyone wishes to have possession of that new invention ! Here comes the role of protecting your invention and avoid falling in false hands !

Patents and copyrights make this protection feasible for long period of time ! Like a researcher completes his research , present it in front of authorized guide or evaluator and then receive the certificate of his academic excellence , he or she can be called as a scientist and then he or she can join any reputed scientific research organization or can become part of big corporate where his inventions will be used as tool for technical upgradation in the organization for which he or she will be offered a good position with high paying emoluments for his lifetime !

After attaining doctorate , researcher can become professor in academic institute and can further improve his research skills by becoming part of scholar's team in that institute where group of like-minded researchers will invent new inventions in typical field of study !

So, how many researchers any industry really needs ? Is there any mathematical relationship between number of regular employees those carry out normal routine task and the special employees who invent new things for the organizational growth prospects ?

If you compare typical requirement of specialized resources in industrial environment a team of 80-100 researcher is a good sign for business size that has nearly 1000-2000 employees !

The team of 80-100 researchers will be split into groups of 8-10 researchers and thus at any time nearly 10 teams will be conducting futuristic research for that organization ! Few teams may be addressing live issues with product after typical service period , few teams may be innovating their product design , few teams may be addressing the untapped market potential , few teams may be working on reduction of product costing , so there will be different targets given to every research team to make reasonable business impact by their valuable contribution ! Some of the team can be specifically assigned a task to bring new advanced machines in the organization with which work can be completed quickly and thus efficiency of operations can be improved !

So , this is how , the life of many researchers revolves around study , inventions , its applications , its financial turnaround , its legal protection and its belonging to society as whole ! More the experienced a researcher is more keen guidance he can provide to his new researchers and thus they can jointly improve the range and quality of their research !

Future Care :

By the time , the early spark and lifespan of researcher is discussed in this preface , it's important to discuss the point of future care taken by scientists and researchers around the world !

Can you ask few questions to yourself ? What will happen , if new natural phenomenon is observed and many people are getting affected from it ? What will happen , if new financial risk is emerged and because of this risk , your money is not remaining in your accounts ? What will happen , if new vehicle is launched in the market and one who purchase it is experiencing sudden fire causing ignitions in its fuel injection system ? What will happen , if people are getting sick when they are travelling from certain countries and when they are joining their near ones , they are also suffering from same illness ?

The first question that comes to mind is why someone is not looking at it ? Who is responsible for this faulty design ? Who will take care of this scientific issue ? Why the deep and extensive study is not carried out to find out its root cause and provide an instant solution ?

This is where the contribution of scientist is noted by the society ! Whenever there is occurrence of some social issues which are beyond the understanding of normal human being , people think about the exceptionally intelligent and brilliant scientist ! It's their ability to co-relate the issue with scientific facts and apply available knowledge to resolve issue in quickest possible time make them a best person in the world for that time !

Scientists and researchers, offers the humanity, care of their future ! Till their invention doesn't become a reality , many people suffer because of live issues present in the society . To reduce this suffering , people likely use available remedies and try to see comfort from it ! As soon as new inventions hit the market,people use it to remove their sufferings !

The role of researchers in this case is to protect people from harmful sufferings ! Take care of upcoming future and make yourself prepared to handle any live situation !

This is why every invention is important for mankind ! Every discovery is important to know the world around us and hence celebrate the life with complete amusement and excitement ! Scientists and researchers are torch bearers who always show the light to people travelling through darkness of ignorance and carelessness ! The path shown by scientist make people aware about available solution to their problems and thus they can get rid of their unbearable suffering !

So , the concluding part of this preface starts with a question that does scientist and researchers enjoy normal human life which has fun, humor, excitement and lot of moments of joy ? Do they party with their friends and dear ones ? Do they enjoy adventurous treks and nostalgic plays ?

The answer to this question is big Yes ! Although the most intelligent people in the society , scientist and researchers have different way of looking at their normal life !

First thing, they enjoy very good financial perks once their invention clicks at right platform ! Once that invention become revolution, they earn in dollars and euros to exceptional potential ! When industry use their invention, they get decently rewarded for that work ! So, the time normal person spends to earn money by carrying out normal duty of 10-12 hours and the time a scientist or researcher spend in their laboratories for innumerable hours of study till that invention become a reality is never comparable ! Scientist are supposed to get excellent financial reward for their utmost dedication, sincerity and sacrifices !

Fun and family moment are quite natural with them and many researchers like to associate with major NGO's and social organizations that make meaningful impact in people's lives ! They are rarely observed with any unhealthy addiction and most of the time their passion for research is seen as a positive addiction ! So, lets dive into such stories of research & scientist which are beyond their inventions !

INDEX

Sr.No.	Description	Page No.
1	Story of Curiosity !	1-12
2	Story of Experiment !	13-24
3	Story of Approach !	25-36
4	Story of will !	37-48
5	Story of patience !	49-60
6	Story of failure !	61-72
7	Story of break & gap !	73-84
8	Story of logic !	85-96
9	Story of intention!	97-108
10	Story of invention !	109-120
11	Guide as coach !	121-132
12	Guide as friend !	133-144
13	Guide as philosopher !	145-156
14	Guide as a question bank !	157-168
15	Guide as a solo viewer !	169-180

INDEX

Sr.No.	Description	Page No.
16	Guide as a reliever !	181-192
17	Guide as a simplifier !	193-204
18	Guide as a sophisticator !	205-216
19	Guide as a presenter !	217-228
20	Guide as a genuine citizen !	229-240
21	Path breaking research !	241-252
22	Impactful research !	253-264
23	Research for Defense !	265-276
24	Research for Business !	277-288
25	Research for humanity !	289-300
26	Research for Agriculture !	301-312
27	Research for good life !	313-324
28	Research for liberation !	325-336
29	Research for values !	337-348
30	Research for advancement !	349-360

LET'S LEARN !

(A-XII)

STEP 1: STORY OF CURIOSITY

" But does scientific knowledge immediately reach to primary school level ? The answer of this question is known to everyone and its big No ! "

Image Courtesy: Karusel , Pixabay.com

1.1 Introduction :

Hello Friends ,

Welcome to the first step of this book ! In this step , we are going to see the role of curiosity in creating a scientist out of a normal human being ! So, to understand this phenomenal transformation , let's see how curiosity generates and works to develop the person thoroughly!

1.2 Questioning Ability :

You may have seen many kids who constantly ask questions to their parents and dear ones ! This is nothing but presence of curiosity ! This is present in every kid to same extent ! Some kids ask so many questions and wish to update themselves with respect to nearby surrounding !

The answers of these questions give utmost satisfaction to the person asking questions because his ignorance about that question get cleared and he receives total knowledge about that question ! Next time ,when he sees same phenomenon , the stored and recorded answer in his memory hints him about that phenomenon and thus he can easily co-relate what is happening around him and why it is happening around him ! This is what the basic scientific research is !

Courtesy : BTP, Pixabay

So , the kid become young and start taking education from premier institute or school ! Inside school , he gets acquainted with typical syllabus of that standard or class and one by one chapters are taught to clarify the surrounding around us !

The surrounding can be about language we speak , the society customs which we accept and cherish , the history of civilization , the geographical phenomenon around us , the mathematics of numbers , general science which is observed in day-to-day life !

With this study of surrounding , without asking questions , students get the readymade proven knowledge which is the result of hundred years of scientific research and development !

But does scientific knowledge immediately reach to primary school level ? The answer of this question is known to everyone and its big No !

1.3 Research to University Curriculum :

Scientific research is first shared within research groups and scientific organizations where research will be presented and scholars will evaluate and analyze its usefulness for society ! Depending upon the usefulness , the research will be either accepted , discussed for more clarification or prohibited for human and animal use considering its risk proportion is more compared to anticipated benefits !

The accepted research will become part of universal publication and it will be published in international journals and scientific societies ! Researchers will get associated with industry to commercialize and scale up the research ! The products will be made and it will be sold to people for their intended applications !

With use of products , people will get to know about the advanced technology and when this technology will become conversant with each other , the space in international publication will become part of post graduate or graduate level curriculum where students will learn about recent advancement !

If you see any curriculum of degree or post graduate courses , the syllabus is generally changed after 5 years ! Why ? The research and development happen with massive pace and because of this the earlier curriculum become obsolete and hence you need to update the syllabus in line with scientific advancements !

1.4 Research to Industry orientation :

What will happen if out of 7 universities in state 2 universities update syllabus after 4 years and other 5 universities update syllabus after 10 years ? The universities which have updated their syllabus after four years will receive recent insights about latest development and employment opportunities linked with that research ! If the research field is extremely advanced , a separate curriculum and branch of study will be created and lessons of that branch will be taught in universities and affiliated colleges ! The colleges also need to show keen interest in starting a new advance branch that will separate themselves from rest of the colleges ! Looking at the employment and future advancement potential , brilliant students will take admission to this new branch and hence after education , the students will contribute to great level to make society easier to live & excel!

For other 5 universities , the syllabus is going to remain same for next 10 years and hence students will receive same lessons even though the environment around

Courtesy:David,Pixabay

them is changing rapidly !

When such pass out students will go for any job interview where the advanced and basic knowledge both are judged equally , these students will clear basic knowledge stage successfully but when questions on advancements in the field will be asked , they will not able to answer completely !

In their comparison , students of earlier 2 universities where syllabus is updated as per scientific research and development , they will not only answer the questions asked but they will also present the pilot project done in those field ! This will immediately add special value to their interview skills and thus they will get upper hand for final selection ! For the presentation of their project work , interviewer will grasp the hidden potential in them and he will offer great compensation package so that they will contribute whole heartedly !

This is how updates of syllabus along with scientific research improves the industrial dynamics ! Industry has to remain updated about various scientific developments taking place around ! If they are not prepared for these advancements , there is always chances to get thrown out of the business from their competitors ! Hence industry is always in need of talent who have researched something new and useful for business ! For such candidate , compensation or perks is not an issue for industry !

So , here we can directly see the interdependent linkage of scientific research with industry to provide new jobs and academics to provide latest knowledge and methodology !

1.5 Stakeholders of research :

Courtesy: BTP ,Pixabay

So , how many people need to stay curious in this happening ? Only scientist ? – No , Only Industrialist ? – No , only academicians ? – No , only students ? – No , only parents ? – No ,

Only education minister ? -No , Only Prime minister ? – No ! Then who ? Who's who ?

Friends , if you want to stay updated about the recent scientific research & development , every individual need to remain aware about these happening ! If you remain aware about these changes , you will swiftly adjust to it and will make a way out of it to stay in line with Fastly changing world ! If you are not aware about the changes , someone other who has accepted that change and developed himself will get that spot and live his life comfortably !

So, the major contribution of scientific research and development is to live life happily ! If science was not there nor curiosity was there , people could not have vaccines for deadly diseases , vehicles for transportation , medicines for longevity and machines for carrying out huge work ! It's the field of science which has tremendously contributed to human society !

On the other hand, inventions also have some negative effect on surrounding and for that point also research is carried out to reduce harmful effect of surrounding and keep planet earth safe to live ! These experiments will become part of future environment monitoring and restoring mechanism ! The task is mammoth but as we seen , with contribution of every individual this will make possible in nearby future !

1.6 Culture of Curiosity :

So , this opening introductory step of the book reveals the need of creation of a curious society as whole ! The society where anyone can ask valid or invalid questions and get the respective knowledge so that after understanding both pros and cons , one can choose the right option wisely ! This will certainly make decision making easy and correct decisions will provide better results !

Curiosity to invention is a big story indeed and, in this journey, scientist has to put their heart and soul to derive the equations which was never known to anyone before ! This is why every invention is so special as it indicates the efforts beyond expectations ! In the next step , we are going to see , the story of experiment and how they inspire to scientists ! ✶✶✶

MULTIPLE CHOICE QUESTIONS

1) As a scientist , how much curiosity is desired to complete research ?

A) Till the breaktrough is achieved .
B) Till the research is sponsored .
C) Till the research is published .
D) Till the research is sold .

2) Curiosity is important because

A) It gives knowledge .
B) It gives sense of belonging to nature .
C) It helps to protect the nature from harms.
D) All of the above .

3) If you are not curious enough to become a scientist , what will happen ?

A) You will not get clarity about science .

B) It will take lots of time to learn key concepts.
C) The research will halt in between because of lack of persistence and self-drive.
D) All of the above.

4) Which of the following question best describes curiosity !

A) Earth is spherical !
B) How spherical the earth is !
C) Why earth is spherical ?
D) Earth is not spherical !

5) About which of the recent invention, you are more curious ?

A) One that has saved many lives of people.
B) One that has saved process cost significantly.
C) One which has broadened profit margin.
D) One that has enhanced comfort level and ease of doing normal work !

STEP 2 : STORY OF EXPERIMENT

"Once the aim of experimentation is determined, then researchers next goal is to look into the type of reaction which will take place during that experiment!"

Image Courtesy: Click Free Vector , Pixabay.com

2.1 Introduction :

How do you know, sun is hot and ice is cold ?

How do you know, water possess three forms ?

How do you know gases can be toxic ?

The answers of these questions cannot be told until you actually carry out a close interaction with these reactants ! You cannot reach up to sun directly but on the basis of sunrays reaching earth in various seasons you can check the temperature of the sunrays and thus you develop understanding of summer and winter season ! You see, temperature is high in the summer while in winter it is comparatively lesser !

To check all states of water, you have to heat it, cool it, to see how the water vapour and ice bar takes its shape ! Just looking at water, you cannot observe these three forms !

To test the toxicity of gas, you can study the elemental composition of gas, you can study already known elements which are known to be toxic and you may try some tests on animals whose body pattern resembles with those of human ! Accordingly, you will note the toxicity level of different gases ! In some of the natural regions , toxic gases keep developing because of reaction of various elements presents there and when human enters that field , he suffers from different pains ! Depending upon the symptoms of gas inhalation , its toxicity is guessed and later sufficient care is taken to wear protective wear before entering such dangerous toxic regions !

So , as a researcher , it is your first and foremost encounter with risks associated with every study ! Till the time you are just studying the known facts , gathering available literature's information , understanding previous researchers work done on given subject , things look under control ! But when you have to experiment something which is yet not known to anybody , that is where the prerequisite of experimental genius comes into place !

2.2 Basic Idea of Experiments in Research :

Every experiment has some type of aim of experimentation ! The aim is intention behind carrying out the experimental work ! The purpose for which things are mixed with each other or interacted with each other ! Once the aim of experimentation is determined , then researchers next goal is to look into the type of reaction which will take place during that experiment !

As we know, every experiment has some type of action with all participants and hence these participants are expected to react with each other in typical way ! The aim of experiment is to find out the rate of reaction , activity level of reaction, presence of reaction products and byproducts , the presence of residues and precipitates ! The indication of reaction which shows reactant are mixed properly !

Courtesy : OCA, Pixabay

The reactions carried out during research can be well calculated or they can be random ! When you know typical reaction formula , you have to simply collect the material and weigh it on weighing scale and then mix it as per given sequence ! As soon as you start mixing elements with each other , they react with each other and exchange their energy with each other in the form of anion and cation ! The resultant reaction can be exothermic or endothermic depending upon the net thermodynamics happened in that reaction ! With scientific proven models you have to note the energy of reaction and note in reaction record , so that whenever you mix the elements in given proportion , you will get the same result !

So , if we consider typical petrochemical reaction taken place in automobiles , the energy of fission after sparking create heat to such an extent that it can drive four stroke or two stroke engines by doing work required to create mobility for vehicle !

In typical steam engine of locomotives , the prime mover uses the energy generated after burning of coal to activate the engine mechanism by which engine become movable and along with its own weight it also pulls other bogies attached to it !

In typical hydroelectric power plant , the static energy of water is converted into kinetic energy and this huge force activate turbine mechanism by which electric generators generate electric energy and same is transmitted to several residential and commercial locations through chain of substations !

2.3 Essentials of Experimenting :

Any experiment is a process of intellectual excellence which is carried out to observe the inter-relationship between various natural phenomenon , natural material and natural mysteries ! The mind of scientist always uses available material to develop method of experimentation and they measure the result of experiments ! The apparatus of machines used for experiment act as an enabler because of which particular scientific change become visible !

So , these 5 Ms of management are essentials of every experiment ! These 5M's are Man (Team of scientist) , Material (All reactants) Method (Combination of scientific processes) , Measurement (Quantum of reaction product and reactants used) and Machines (Apparatus , mixers , dryers , heaters , coolers) which are used for experimentation !

2.4 Chances of Failures :

Research and development are extremely challenging field and to gain expertise in this field you need to aware about fundamental science , advanced experimentation techniques , new machines and deep understanding of environment around us !

Courtesy : CFV, Pixabay

A researchers mind always need to think before taking a single action step ! This is because , they are directly dealing with natural phenomenon and unless you are pretty sure about the specific consequences , you are supposed to not take the risk of experimentation !

If the material is flammable , you cannot ignite it without having proper safety manifolds

! If the material is radioactive , you cannot touch it without using radiation safety equipment's and PPE ! If the material is good absorbent , you cannot keep it in open places by which it may absorb undesired foreign elements ! If the material is highly flowable , you need to store in closed container ! If you not take safety steps , then as per their natural behavior , they will show the respective properties and to control their behavior you have to use control measures opposite to their natural behavior ! If the material catches fire ,you have to use fire extinguishers and you have to ensure , the reaction is favorable to stop the fire !

2.5 Chances of Success :

Based on number of attempts of trials and errors and uses of more than one theoretical and practical action plans , experiments meet with success finally ! This success is used to develop and confirm the findings of experiments !

The success of experiments also depends upon whether you are carrying out experiment individually or with the team of fellow researchers ! The advantage of carrying out experiment single handedly is you can work as per your own conscience of fundamentals and there could be very little scope of different opinions regarding accepting a particular action plan ! Here one can use absolute freedom to apply their scientific thinking behind that experiment and see the possible result ! The credit of success as well as failure has to own personally !

The advantage of carrying out experiments with team of researchers is you get different views on same subject and team can handle different issues associated with the experiment with collective intelligence ! You can thoroughly participate in between different brain storming required before experimentation and people can share the experience of diverse fields to add value to your experiment ! The success as well as failure of the experiment is jointly owned by the team of researchers and every effort is made to achieve required finding of those experiments !

2.6 The most suitable path for experiment :

Courtesy: CFV, Pixabay

Experiments can be carried out by more than one method . The aim of researcher is to find whether results remain same or they haphazardly changes to extreme variation !

When more than one method of performing an experiment is find out , researchers get more convenience to reduce the required speed of reaction and thus they save considerable process time required to find out inference !

The experiment must be successful and it must complete in least possible time ! This is expected to achieve ease of repeat experimentation ! When you are experimenting on larger sample size , every individual experiment long time consume more time for complete sample size testing , that alternatively increases time required for inference making ! Herewith , we are moving towards next step of approach of scientific research behind every invention ! ⊛⊛⊛

MULTIPLE CHOICE QUESTIONS

1) Which of the following experiment will take more time to finish ?

A) One which has numerous complex reactions.
B) One which deals with high safety risks.
C) One which is carried out first time .
D) All of the above

2) Performing experiment with team of researchers has advantage of

A) Collective Learning .
B) Team conscience generation .
C) Open and collaborative atmosphere .
D) All of the above .

3) Which of the following concept need to be very very clear before carrying out experiments ?

A) Aim of the experiement.
B) Method of carrying out experiment.
C) Previous failures of other researchers.
D) Material wastage during experiment.

4) **The best part of every successful experiment is**

A) Getting proven clarity about previous assumptions.
B) Finding reaction formula for more accuracy
C) Commercialization in future opportunities
D) All of the above.

5) **The process of identifying names of chemicals with the help of stickers is known as**

A) Labelling
B) Stamping
C) Embossing
D) Pasting

STEP 3 : STORY OF APPROACH

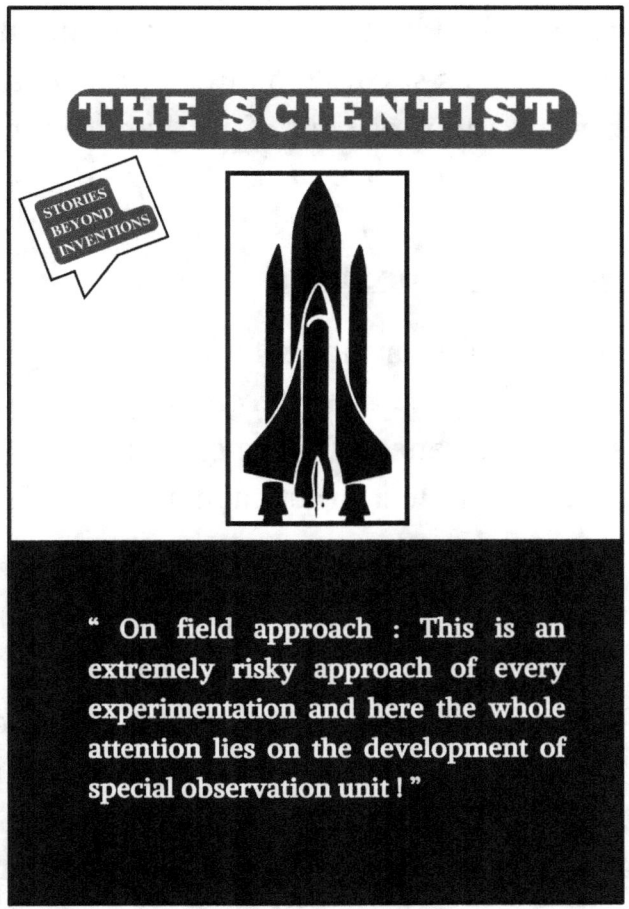

Image Courtesy: Click Free Vector , Pixabay.com

3.1 Introduction :

Once the aim of experiment is determined , after building curiosity of research in that experiment , the next step of research is approach to tackle the experimentation . In this discussion , let's say how many approaches help to derive an inference out of any experiment !

3.2 Modular Approach :

Modular approach refers to building a test model and carry out different experiments to derive the inference of those experiments ! This is first and foremost practical experience to any researcher where he can implement experiment step, on model created or designed by them !

Suppose , a researcher has designed a bike that can provide power transmission through combination of solar energy & electric energy ! Can this possible ? Yes , this is what the research is ! In research , you have to challenge the status quo and find out the possibility of success !

There are two requirements before this trial ! You need to know how much power is required to start the bike engine and keep it working as per your requirement ! In traditional fossil fuel operated bikes , the fuel tank has specific capacity ! Once the spark is generated and you start the vehicle , fuel consumption takes place , as you go ahead , energy conversion takes place to give rise to mechanical energy required for motion of the vehicle !

Courtesy: OCA , Pixabay

In electric bike , same mechanical energy is received after conversion of electrical energy stored in batteries ! When battery is fully charged ; vehicle starts just after pressing start switch !

When battery is low , the energy required to continue movement decreases and as a result of which vehicle stops ! So, there is specific energy requirement to start and run the vehicle !

Solar energy is abundance in nature and in near future someone can develop a model where solar panels will receive the heat energy , will give that input to create electric energy and based on this electric charging facility , people will charge their vehicles ! Can this possible , certainly , because as per universal law of conservation of energy , " Energy cannot be created , cannot be destroyed , it can be converted from one form to another ! "

We have earlier seen ; the static energy of water is converted into kinetic energy by which turbines are moved to generate electricity ! In the case of solar energy , amount of heat available between peak sunlight time , probably between 11.30 to 3.30 pm , one can easily convert this solar energy into electric energy and same can be made available to vehicle charging !

Now , when this modular approach will work successfully , just imagine how easy people live will become ! The environment will become clean because of less emission of pollutants and the cost of fuel will also fairly drop down !

Now , as a researcher , he has to approach this invention with different trials on his model ! Initially he may experience issues in availability of proper sunlight that can activate electricity generation model. Then he will note, within this particular time of the season ample sunlight is available that can convert energy equivalent to half liter or one liter of fuel average distance covered ! With this study, he can achieve partial success and later can research further on electricity model with which he can quickly carry out conversion. In this approach, he may experience that his charging capacity is improved and it almost become double ! So, now the vehicle can travel distance equivalent to 1 liter fuel average ! In next step, they will change the size of solar plant and see how the electric energy capacity increases ! When positive results will be seen, then they will carry out experiment to generate electric energy from solar energy equivalent to daily 5 liters of fuel consumption ! With this success, they can present the invention and they can claim 'First Sol-electric – Solar + Electric bike is arrived that can give mileage of 60 Kmpl (Petrol) or 30 Kmpl (Diesel) !

With this invention, people will get effective option for fossil fuel and huge burden on national economy about rising fuel prices will get reduced ! This in turn will reduce prices of other items which were risen because of increase in transportation cost ! So , one scientific invention can transform your economy and provide you an excellent opportunity to grow your export business and thus improve your foreign exchange rates !

3.3 Theoretical Study Based Approach :

In this approach , in contrast to modular approach , comparatively lesser practical trials are taken and main focus is kept on finding inter-relevance of number of scientific concepts !

To simplify this understanding , let's consider a disease treatment done by a medical practitioner with the help of available medicines and the disease treatment done by a medical surgeon by carrying out required surgery !

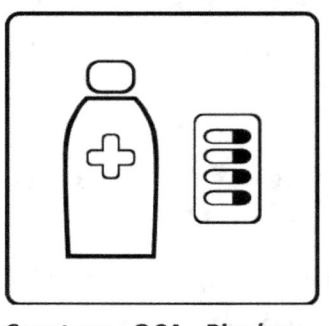

Courtesy : OCA , Pixabay

In first practice , causes of disease , lifestyle of patient , symptoms of disease are studied and medicines are developed to either eliminate the root causes of the disease or to keep themselves under normally acceptable levels ! For this you have to consume medicines as per prescription for lifetime ! Your fear of surgery and possible risks get reduced with this approach ! Whereas in case of surgery , the research taken in surgical domain is utilized which can be available in the form of additional qualification of surgeon , availability of new technique , improvement in technology , development of new procedure , availability of artificial parts and their suitable body acceptance ! Based on such type of research , patient may choose to get operated instead of taking lifelong medicines ! Secondly , because of reduction of post-surgery risks , there is no fear for anything ! This also helps a patient to accept the surgical path of comparatively quick relief !

So, the development of various drugs and vaccines can come under theoretical approach of research . Here the aim of experiments is to derive a scientific formula which can reduce disease symptoms and make patient feel nice and comfortable !

3.4 Random Approach :

Can experiments be done with random approach ? Yes , why not ?

When any first of kind trouble is noted , researchers first try to identify its behavior with deep and constant observation and monitoring ! As there is no specific laid down path is available at that moment , they choose their random experience to deal with the case ! The team of researchers will choose most suitable stand at first instance and then they will note the change between the behavior of trouble ! Secondly , they will take a stand , which they know it will never work , but still, they will take a chance to see the behavior ! Now , if they receive expected result , then it indicate , their assumptions are correct ! If contrast observations are noted , then they understand ,this trouble doesn't have any specific orientation and hence its control solutions need to be available on both side of possibilities ! Thirdly , they may note how much positive and how much negative changes can happen in that behavior ! This sets positive and negative range of that trouble and thus a random trouble is caught inside a specific range which can be controlled by providing available solution or by providing new solution !

In such random experimentation, defining the type of trouble is very much important. In same type of trouble, you see same symptoms ! In different type of troubles, you see different symptoms ! So, your cure plan will be different for different types of troubles ! And when you see ,altogether new type of trouble, you have to recognize it and identify it as a new species of trouble ! Your total approach to control this trouble will be mixed of new solutions and part of proven solutions !

3.5 On field approach :

This is extremely risky approach of every experimentation and here the whole attention lies on development of special observation unit !

To explain with the example, understand various research carried out in special space missions ! It's the best example of ' on field research ' !

In this approach, you have to develop observation unit in such a way that it can go

Courtesy : Inesjimenex, Pixabay

there safely , collect required information , send that information to you and either it can stay there or it can return as per your planning !

The observation unit can be manned or unmanned ! With increase in risk of addition of human , researchers have to develop all necessary safety measures with which man can complete necessary observation period !

If first such type of trial become successful , next trial is done with additional capacity enhancement and this process goes on !

Same type of experiments can be seen in uses of nuclear power where on field observations are carried out to see its overall effectiveness and disadvantages !

In next chapter , we are stepping into story of will of a researcher ! The impeccable difference maker for every researcher ! ✲✲✲

MULTIPLE CHOICE QUESTIONS

1) Identify the approach of research in carrying out experiment of launch of space vehicle !

A) Laboratory Based
B) Industry Based
C) Academy based
D) On field based

2) In modular approach of experimentation , researchers develop a design that can

A) Take care of system & surrounding .
B) Cost efficient and easy to use .
C) Reproducible and scalable .
D) All of the above

3) Given a chance as entry level researcher to work in research institute that

deals with health sciences , what would be your priority ?

A) Work as per research institute work protocol.
B) Work on most urgent research projects .
C) Read, observe and understand the way of working in research institute for initial time.
D) Both A & C

4) **What will happen if research approach is incorrect ?**

A) Experiment will take more time to derive any inference and it will set incorrect research practice .
B) You may find out more quick solution with incorrect approach . Research demands degree of freedom of approaches until it discovers aim of the experiment !
C) Many inventions are discovered while carrying out experiment with different approach !
D) All of the above !

STEP 4 : STORY OF WILL

" When there is collective willpower , many people can contribute to same issue and with this involvement , issue looks simple ! Once the issue become simple , possibilities of its resolution get increased ! "

Image Courtesy: Geralt , Pixabay.com

4.1 Introduction :

Many researchers need to love this quote before starting their regular research work , the quote is ,' **where there is will , there is way !** ' How meaningful and inspiring this quote is for those people who are on the journey of scientific research and advancement !

What will happen if curiosity is there but will power is not there ? People will start the research with enthusiasm necessary and as they keep progressing ahead , they will develop more interest about that research . A stage will come where researchers will get stuck in between and they are feeling like this is the end of this research . We have to return back because the road ahead is not safe for us nor for any human being !

This is the point of research phase where your willpower makes wonders ! You stick there , research further by taking some known risks and then observe the light at the end of tunnel ! Because of this risk-taking willpower , you have resolved the mystery present in the experiment and now you can document it to observe it again and again !

4.2 Series of Experiments & Willpower :

Sometimes as a researcher , you have to carry out series of experiments ! Result of one experiment is input of second one and result of second experiment is input for third one and likewise ! It's just like solving a mathematical sum where you have to calculate various entities to derive final answer of the sum !

In all this study , you have to keep one simple thing in mind , may heaven fall, I am not going to return back till I reach up to a satisfactory conclusion of this series of experiments ! The ultimate dedication of a research fellow is always observed with this never ending quest for new inventions that can make human life easier !

In such series of experiments , first few experiments will give required result easily but as you progress to more complex experiments , you need to study and observe each parameter carefully to observe the pin-point effect because of changes in series variables !

This is where fact finding habit of researcher comes into play ! Towards every experiment , they have to observe with scientific vision and they have to analyze how this series of experiment will find useful observation for rest of the experiments !

Success in experiments comes after many trials and errors ! You arrange things required for trial, decide their sequence of experimentation and suddenly one element start functioning erratically !

In some experiments , you are almost near to final conclusion and before confirming the final result , you just check some of the parameters and you observe there is serious defect in main control unit !

In some experiments , you carry out first trial correctly but while taking second trial ,you don't get desired results and observe major shift in the observations !

Such type of high variation in research is quite common thing and hence every researcher has to adapt a ' kaizen approach ' that reflects scope for continuous development ! Learning never stops and hence researchers need a strong willpower to learn why something is not happening as per desired expectations !

4.3 Collective Willpower of researchers :

They say , ' when its heavy , lift with both hands !' What is the hidden meaning of this quote ? This quote talks about the collective willpower of researchers ! Every research team has some of the gem of research field. The team consist of senior scientist who has fair amount of research experience and they have almost every high-profile connection where they can get immediate access for decision making !

Second stream consist of talented scientists who just hate to give up ! They have enormous willpower to research new things and they never stop till the expected results are received ! Time is not a factor for them ! They simply stretch themselves in the field of research to such an extent that they forget to take food on time , sleep is almost shortened and they have extremely low degree of social life ! That only experiment develop passion inside them to keep strong faith in what they are doing and how it will become a massive invention once its results are achieved !

Once the research is completed , they celebrate it in style along with their peers and then swiftly move to next research !

Courtesy: Jalexiscv,Pixabay

The last level consists of talented new research entrants ! As far as their field experience is concerned its least than every other senior member but as far as their knowledge bandwidth is concerned , they are almost equal to a research fellow who has 5-10 years of experience ! Does it happen really like this ? Yes ! This is the pace of research is !

The new research entrants have readily studied the literature which research was taken place almost ten to twenty years back ! So , once they complete literature study of that research , they can carry out respective experiments and thus acquire knowledge of those fields !

Now whatever they have to add is next step of that research ! So ,when these knowledgeable researchers join hands with experienced researchers, after duration of two - three years, they gel up with the core team and thus the accumulation of collective willpower is witnessed !

When there is collective willpower, many people can contribute to same issue and with this involvement, issue looks simple ! Once the issue become simple, possibilities of its resolution get increased ! The responsibilities are shared with involved researchers and study is carried out further ! Every researcher presents their work in collective team presentation and thus everyone gets the idea about whole work being done !

When such type of time-based research reviews are carried out, after every review, team observe improvement in common understanding of the issue and thus they come closer to solution !

With almost 90% progress of the research, finally the breakthrough is achieved by either of an entry level researcher or by senior researcher and that study become complete for that scope of research !

The documents are created, proto types are preserved and the scholarly articles are published to tell the world, that ' Hello World, we have arrived ! We have successfully and collectively cracked the mystery behind this experiment ! Now, this invention is yours ! '

4.4 Willpower in personal emergencies :

The life of researcher is not an easy life ! You are always sitted on first chair of the class and you have to be very much attentive to surrounding around ! It's your prime responsibility to note changes happening in surrounding and find out whether they are desirable for society or undesirable for society ! Based on your lawful observations, you have to either support that change or express strict prohibition on that change !

As you know the structure of society, if you oppose to something, you have to face oppositions views and comments which can be verbal test of your cool and calm composure ! Still, you have to insist that your opposite view has a direct connection to risks associated for humanity or general well-being and in no case , this change should be accepted by society !

Apart from this change, researcher has his own personal life which can be good -better - excellent or which can be little bit challenging and struggling ! Because money cannot solve all problems ! Few problems need compassion , understanding and time !

Courtesy: Banger , Pixabay

So, a researcher facing the heat and cold of personal and professional urgencies is like a master falled in death trap ! On one side , he has to deliver to people's expectations and on other side , he has to take care of his small little world ! Both things are extremely important for him and for this he is committed with heart and soul !

Where the strength comes to deal with such a challenging lifestyle ? This strength comes from willpower to experiment and keep experimenting ! Constant learning approach that can be regarding personal life events or it can be professional milestones , they develop required willpower inside you to cope with the pressure situations and handle them softly yet decisively !

Some researchers may face several attacks of chronic depression and anxiety since the waiting time required for success of experiment could range from one year to a decade ! How one can remain so calm and composed till such a huge waiting time ? But this is what the profession of researcher is ! At the end of a decade or more than that , you achieve your desired dream for which you spend a bright and energetic decade of your life and this is giving you returns you deserved ! This is happened only because of your will power and belief in your own thought process and perfect actions !

4.5 Willpower in boom !

Do you know , sometimes in life , researchers get all round support and appreciation for their contribution to society and they become Page 1 interest for every news line

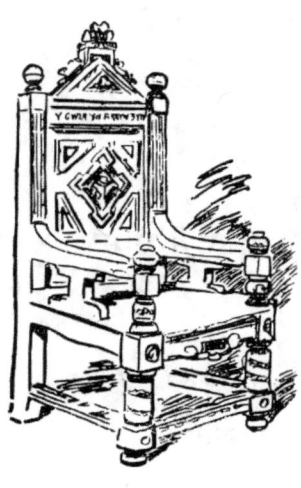

Courtesy:OCA,Pixabay

! This is the phase of boom for every researcher ! In this phase , they have to receive the praise and appreciation in true faith and spirit and gently need to move towards next experimentation ! This move also need great willpower of concentration and constant learning !

Ultimately , researchers need to possess a great proportion of element of rationality in their thinking and living to make it more sensible as well as responsible in long run of life !

How to increase willpower of a person is a million-dollar question and its importance is only felt when a person faces several failures in his life because of the reasons for which he may not be solely accountable but they have to accept the final call of the experimentation ! This is where indomitable spirit of researcher does the trick and bring his bright mind back on track with great willpower ! In next step , we are dealing with story of patience ! ✸✸✸

MULTIPLE CHOICE QUESTIONS

1) **Tell the best example of willpower in given situations :**

   A) Research Success after 15 years !
   B) Research success after 156 trials !
   C) Research success of team of 241 researchers!
   D) All of the above !

2) **What will you prefer to do in following situations ?**

   A) Repeat the experiment after 7th consecutive failure !
   B) Leave the experiment after 99th trial which was approximately 89% successful!
   C) Hand over the experiment to junior after spending 12 years in research !
   D) Will stop reading research papers when count reaches 19084 papers , as you know almost everything by now !

STEP 5 : STORY OF PATIENCE

Image Courtesy: Geralt , Pixabay.com

5.1 Introduction :

Friends ,

Who can understand value of patience better than researchers and scientists ! They are putting their lifetime to invent new things which are present in this very nature but yet not known to all of us ! Finding such natural phenomenon and linking that phenomenon to development aspect of humanity is such a noble task and many times Nobel prize of high acclamation is given to fantastic scientists whose research changes the earlier perspective to look around the nature !

In this discussion , we are going to see importance of patience in research work and how to build patience in most impatient situations !

5.2 Most impatient situations in research :

As you know , entering and touching the research domain means entering into unknown territory with the help of basic intellectual ability present with human being ! You have to start your work with some prior assumptions and understandings and then slowly test the parameters present in given environment to cross check with those assumptions ! The degree to which your prior assumptions match to observed natural phenomenon , create moment of success but the deviation to prior assumption again need to think about different perspective of that phenomenon and hence you have to start the research with new perspective !

Courtesy: Prawny,Pixabay

So, just understand ,what can be status of mind when you come so close to an observation and when you are sure about drawing a particular inference , suddenly the phenomenon suggests that this is not as you are expecting ! There is difference in your understanding and for which you have to further analyze the behavior to accurately record the pattern of this phenomenon !

So, how many impatient situations are normally occurs during a research work ?

Well, if previous research experience of great scientist is concerned , some scientists failed almost 500-700 times during their research work , some spent nearly 15-20 years while researching on only one phenomenon , some researchers while carrying out research of one subject , observed presence of different natural phenomenon and this has changed the direction and scope of research !

Apparatus failure , reinstallation of connections , capacity mismatch to expected experiment need , occurrence of defects during trials , safety hazards during trials , social stand during research , prohibition on research from international social welfare communities , these types of hurdles are always part of every researcher's life and work ! To deal with these hurdles , they need great amount of patience and they need to convince their stand in such a way that people believe it totally !

What happen when a new model is developed , sold and after sales many quality issues are noted ! Firms has to call back those models to correct those quality issues by fair amount of research and they have to ensure new models are equipped with more strength to eliminate observed quality issues ! This phase really makes any inventor impatient because this aspect was never thought while carrying out research as such !

Research funding and policy makers back up is always make any researcher impatient ! Money is required to carry out research . Your domestic research unit may have some equipment's and may not have some equipment's ! For this resource deficit , you need to associate with external facilities and need to sign interpersonal professional co-operation agreement to carry out research in their area !

Again , if you have to keep your research confidential , you need to make non-disclosure agreement with that firm ! Here firm may co-operate with you or they may deny signing on such type of deals !

Policy makers support is essential to give morale boost to researcher ! Considering the huge financial investment in research work and possible chances of failure , it's not easy job to support every research ! This is because , every research cannot become successful ! Till the breakthrough invention happen , several experiments eat up your research funds and they derive little information ! This situation can make a scientist and his team impatient for long time ! Here the role of policy makers comes into play ! As a governing body of that nation, they have to encourage researchers by going beyond normal call of duty ! See, every country , nowadays , has understood the importance of research and to stay relevant to scientific advancements , policy makers need to support research work strategically, financially and most important - ethically !

When you know three of your five neighbor countries are carrying out research in particular field that can influence future progress in employment generation , agricultural output improvement , defense related improvement or health related improvement , your nation also need to participate in such type of research wave to stay with the other nations ! Ultimately , when your scientists will crack the secret of that invention , it's your country which is going to receive the benefit of that invention ! If you commercialize that invention , it's your country which will supply such products to other countries and thus your financial conditions will improve !

This is why policy makers interest and support is essential in such type of research programme ! If this support is not available , scientist and team of scientist can do what is possible with available resources and they will figure out the development plan which can be done if their funding and other ethical matters are supported by governing organizations or policy makers !

5.3 Strategies to develop research level patience :

It can be fairly quoted that a twenty - twenty cricketer can play many big shots in quickest possible time but a test batsman bat for almost three days' time to collect nearly equal or double runs ! So , what was started earlier ? Test cricket or one day's or twenty -twenty ? Off course everyone knows , its test match cricket that introduced thrill of cricket to people by playing two innings from each side in five days available time !

Courtesy: Jainiox, Pixabay

Job of researchers is not less than any test cricketer ! As a test cricketer , they need to be technically strong , they need to be methodical , they need to understand the merit of the delivery and they must play according to that merit , there is no need to do any hurry if things are moving following a particular pattern , you have to understand the weather condition and change your game according to most suitable run making stands ! You have to remember that on a given pitch , you have to keep your wicket intact , you have to also ensure you make a formidable total on the board , you have to ensure you defend bouncing deliveries and you need to ensure that some deliveries are there to left as it is when they are bowled !

All that matters is reliable and progressive partnership with your fellow researchers and making sure that research work is facing right wind and using it for improvement of society !

So, to develop required research patience, researchers have to improve their reading aptitude, they have to improve their analytical aptitude, logical reasoning, out of the box thinking, in fact this is most desired pre requisite of every researcher's natural research instinct, they need to stay curious and focused on their research goal, they need to be light hearted and every time they can't stay serious !

They need to understand, sometime things take their own time and they need to stay patient and humble and genuinely march ahead for further work ! One minute of impatient feeling can disturb your several years of research and hence though heaven falls, you have to hold your nerves and need to take decisions which outplays the test of time !

They need to be fearless about performing experiments but at the same time they need to be cautious about external surrounding which has no direct connection to their experimentation. In no way their experiment could harm the innocent external surrounding ! Perhaps, this is the most important demarcation line in the field of research ! Until it's not trialed and tested , inventions or research work need to keep secret ! No one should make false use of these phenomenon !

5.4 Networking for research patience :

It is a well-known fact that , during research , you may need to discuss the progress with your peers , research guide and other fellow researchers ! The importance of research networking is observed when you need typical service and it is not available in your work facility . If you ask about it in your research network , you can easily get that service by following typical protocol of that facility ! If some other research team doesn't have a typical resource and if your facility has that resource and it is lying in idle state , you can always provide that resource to these researchers ! Such sharing converts impatient moments into moment of comfort ! In next step , we are discussing story of failure ! ✳✳✳

Courtesy: OCA,Pixabay

MULTIPLE CHOICE QUESTIONS

1) **Which of the following situation defines the best patience level of a researcher ?**

   A) Gathering 1000 observations for selected sample study !
   B) Calculations of 1000 observations to put it on defining mathematical relationship .
   C) Setting equation of graph and finalizing the experimental analysis !
   D) Taking 1000 random observations from sample size of 10 Lakh !

2) **Identify the most impatient situation from given list of experiments ?**

   A) The required result value is missing by 0.1% after 25 trials !
   B) After 1 Lakh liter consumption of test fluid , test result is sleeping out of range by 100 ml
   C) Trend of repetition of result is not constant after 5 experiments and hence 1 year

invested in these 5 experiments is going in vain !
D) All of the above !

3) How you will approach to your research network for sharing laboratory resources for an urgent experiment ?

A) By filling applicable system requisition on dedicated website !
B) By calling on Phone .
C) By meeting through video calling app .
D) By sending an informal e-mail !

4) Who of the following possesses best patience level out of given professions ?

A) A high-profile diplomat dealing with one international conflict !
B) A nursery teacher who is teaching newly admitted extremely sensitive 3 kids !
C) A music director who is working on 7 projects at a time !
D) A scientist with 100 patents on his name !

STEP 6 : STORY OF FAILURE

Image Courtesy: Open Clip Art , Pixabay.com

6.1 Introduction :

Can you show a researcher who is afraid to fail ? Can you show a scientist who has not a single time failed in his career ? Can you show new research which is passed all performance test in first instance and repeated for several years later ?

Friends , these are some of the ideal conditions desired in every researcher or scientist ! When you deal with idealism , there is no scope for error and faults , whereas when you deal with practical concept , there is comparable scope for errors and faults ! This is because , this approach supports basic life philosophy which is , ' To err is human ' or only human makes mistakes !

In this discussion , we are going to discuss story of failure and how researchers need to overcome it to continue their research to further level ! So, let's understand failures in research field step by step !

6.2 Approach Failure :

Approach of experimentation is extremely important when you are expecting fair success level within practical time limits ! If you choose incorrect approach , your results are going to differ to considerable extent and you may need to repeat the experiment with different approach all together !

6.3 Process Failure :

Courtesy: CFV, Pixabay

The design of process steps while carrying out experimentation is very very important ! Every process has particular sequence in which desired reactions are bound to happen . If because of mistake or because of overlooking , if process sequence become incorrect , chances of success reduces and chances of errors increases ! As a keen researcher , when you design a process for any experiment , its wise idea to get it reviewed from experienced research guide so that with the help of their prior knowledge they can confirm its logical accuracy before practically mixing the variables ! Once their approval is received , you can go ahead step by step and observe how that process of experiment behaves with interaction of process elements !

6.4 Formulation Failure :

Do you know , the formula derived after any experiment is exact indicator of experiment success ! During experimental formulation , you have to study various samples to check their response to particular scientific phenomenon !

For example , in carrying out difference in ultimate tensile strength result of newly developed alloys , the formula for calculating tensile strength remains same as it is derived after empirical research process and it is not going to change anywhere !

So, with the available formula , you just have to load new samples as per applicable weight and observe their ultimate tensile strength before breaking into two parts ! You also have to note their percentage elongation !

So , if you have tested 5 new alloys of different grades , you will calculate 5 UTS values of individual alloys and thus you will get information about most suitable alloy for your intended application ! When codes and standards were not available , which resource was used to ascertain the UTS values ? Off course , practical testing of these alloys in approved test laboratories and then new alloy development research papers with alloy properties were published in international material journal and magazines !

So, this is where formula plays important role ! If you look at stress theory , it all starts from concept of elasticity , modulus of elasticity , study of stress -strain curve , definition of stress and strain and their formula , effect on material after experiencing different loading conditions , elastic and plastic zones and determination of yield stress and ultimate tensile stress , observation of material failure and formation of neck during failure ! So , when all this study was carried out , then formulae of stress , strain , modulus of elasticity were derived and used for study of metallic objects ! Based on observed strength of the material , specific material is recommended for specific matching application and thus the trend of using right material for right application at right price is started !

With incorrect formulation , one cannot arrive at logical conclusion ! If you study any formula, it indicates the result of reaction has a particular mathematical connection between reactants involved in that reaction ! Some reactants will directly and proportionately interact with each other while some will remain there just as an element ! There will be specific constants applicable to those reactions and these constants never changes after changing the stages of reaction !

Courtesy: CFV, Pixabay

So , the aim of researcher behind any experiment involves setting the relationship between variables and fix elements along with constants of the reaction !

Typical formulation of scientific experimentation can be summarized as follow :

1) Velocity = Distance / Time
2) Voltage = Current x Resistance
3) Circumference = 2 π r , Area of circle = π r²

What is special about π ? It's well-known constant and its value is confirmed as 22/7 or 3.142 !

So , what will happen if someone refer value of π as 23/7 or 20/6 ? Your all answers will produce errors and you will not able to find out right circumference and right area of circle ! Now if you are building an experimental model in which you have to construct a circular shape , because of incorrect constant reference , your dimension will either become more or less than required circumference and hence you need to carry out error correction by either cutting material to correct size or adding a suitable joint ! So , this is why accurate formulation of experimental work is necessary ! That one experiment is going to become base for future mass production !

With this example one can understand , how the calculations of applicable stress on a part are important before deciding its final dimensions ! If you are designing a part , you need to understand typical loading condition and presence of repetitive stresses , you need to understand corrosion and wear allowance and you have to note co-efficient of thermal expansion , you have to note design factor of safety and also add additional allowance for friction if any ! When you account all relevant design factors , then only by using correct formula ,you arrive at required dimensions ! Next task is sizing to those dimensions and making parts with right manufacturing processes ! If your all formulation is correct , the test results will be satisfactory !

6.5 Failure because of conditions :

As we know, scientific experiments are carried out at different atmospheric conditions . You have to monitor performance of your device at different atmospheric conditions to see its rigidity , strength and usefulness ! In typical laboratory environment , what you do , you try to create similar type of artificial environment to study the effect on your device !

So, to study the effect of corrosive environment on your given grade of steel , what you will do , you will put specimen sample in proportionate sample of acidic or basic environment and check response of steel to that solution ! If the rate of corrosion is higher than expected rate calculated after considering minimum design life , the sample will fail earlier than its intended failure time ! If the rate of corrosion is less , sample will not corrode more and will pass the test ! So, when actual job will be put in such corrosive environment , till the laboratory test conditions and actual working environment are matching each other , job will function normally , if major conditional deviation is noted , job will fail before its intended design life ! That is the reason why you need to monitor system and surrounding in totality to avoid early failures of your functional parts !

Courtesy: CDZ, Pixabay

A prudent designers or researcher always try to develop a machine that can perform in harshest environment reliably ! All they need to do is study the environment to full extent and then add required strength to their design ! This ensures, the design is safe and part can function satisfactorily till the operating instructions are obeyed !

6.6 Unknown Failures :

During the early analysis of experimental evaluations, there can be many factors which are not taken into consideration but they are present there ! If they are not taken into consideration, the early hypothesis will orient to incomplete direction and results will be incorrect !

When you search those unknown elements and include in formula as additional measure of variation or accuracy improvement, your failures get reduce ! So, every experiment is just like solving a mathematical puzzle !

You have to find out the tricky arrangement of observed data and just crack the required code which can satisfy the equation and make that unknown element a better known one!

In next step, we are going to discuss story of break & gap which is next step of research and researchers know its importance ! ✹✹✹

MULTIPLE CHOICE QUESTIONS

1) Which of the following is considered as a process failure ?

A) Occurrence of same defect after every trial !
B) Occurrence of defect in one part per ten parts produced !
C) Passing of all 10 parts which are produced in one batch !
D) Passing of 7 parts and repair of 3 parts which are produced in one batch !

2) Which of the following researcher will probably complete his research in time ?

A) Sam has performed research under Mr. Shyam , Mr. Shyam has reputation of carrying out research with most advanced laboratory equipment access in all over the world !
B) Sheela has 10 years of research experience and she has filed 3 patents on her name !

C) Madhu is an industrial expert and he has monitored 20 undergraduate students' projects ! Now Madhu is doing research in his field of interest !
D) Reena and Amar are carrying out research in 90 % completed subject ! The balance work is being researched since last 5 years !

3) **Which of the following conditions are challenging for any research where chances of failure are maximum ?**

A) Carrying out research in subzero temperatures where frequent spells of thunderstorms and ice formation is taking place !
B) Carrying out research in ultra hot zones where average days temperature is nearly 47degree Celsius and average night temperature is 38 degrees Celsius !
C) Carrying out research in risky gaseous environment that has presence of gases like H_2S, CO, CO_2
D) All of the above !

STEP 7 : STORY OF BREAK & GAP

Image Courtesy: Open Clip Art , Pixabay.com

7.1 Introduction :

Hello Friends ,

What do you mean by a break and a gap ? The question looks extremely simple and casual but it has deep meaning when this concept is applied to field of scientific research !

The field of scientific research deals with futuristic developments to tackle the upcoming challenges ! When the whole world is busy in living normal life , scientist's life is always struggling to find out the breakthrough invention that will make human life easy and sustainable in spite of varying degree of natural and man -made challenges !

So , when scientists decide to take a break in their experimentation or when they decide to take a gap in their experimentation , it has very very deep society impact attached to progress of that particular research !

In this step , we are going to see , how research breaks and gaps affect regular life flow of people and their various associations ! Life is seamless , it doesn't wait for anything , you constantly keep growing and in this development time you need to be aware about what's happening in your surrounding ! If you could easily relate yourself to scientific advancements happening in and around your locality , your materialistic progress chances will increase !

7.2 Short Research Breaks :

Courtesy : Artsybee, Pixabay

Carrying out research is extensive time eating activity that one need to do with total caution ! In some of the long experimental studies , you need to stand in front of apparatus or model for several days and night to notice the minute alteration and how it is happening from its start to end !

This observation make yourself tired physically and mentally ! But the outputs you receive after such experiments always inspire you to go ahead for next experiment !

Short breaks are taken before starting new experiment just to feel afresh , collect the learnings of prior experimentation and deciding the roadmap of next level of experiment ! This stand make yourself ready for new trials !

7.3 Medium Level Breaks :

Again , medium level breaks can be co-related to stoppage of some earlier experiments and moving towards starting of new experiment whose result will support completion of earlier experiment !

It means suppose , an experiment A has total 10 steps to complete till final inference and experiment B has total 5 steps to complete and if step no 7 to 10 of experiment A can be done only after getting the inference of experiment B's 5 step completion , then you have to first complete experiment A from step 1 to step 6 , take a medium level break at that stage , go on for experiment B and complete its 5 steps ! Once result of these 5 steps is received , again start the experiment A for balance steps, which are step no 7 to step no 10 ! At the end of step no. 10 , you will get the inference of experiment A !

The above example is given for simple understanding . In full-fledged research work , there are multiple experiments going on in one research team ! Here research co-ordinator has to receive output or inference of completed research and then he has to find out the possibility of further research completion ! It can be like job of a bus depo co-ordinator who check which bus is coming to depo , which bus is leaving shortly , which bus needs fuel refill and which bus needs maintenance ! Thus, research co-ordinator has to manage the inflow and outflow of every experiment happening in that research team ! So , if 5-7 teams of researchers are working , it can happen that 2 teams are taking medium level break and 3-5 teams are busy in balance research ! Once their output will be received , the earlier teams will be engaged and other 2 teams may take medium level breaks ! Because of such breaks , the work streaming become systematic and researchers could release the natural stress that come during research !

Since the research work is purely a brain work and it has comparatively lesser physical exhaustion , you need to keep your mind and brain fresh and active to invent things that really contribute towards progress of society !

Human body limits are known to everyone and scientists are also not different to this phenomenon ! Hence to keep their energy level high and higher , such breaks help them relax for a while and again join the research work !

7.4 Long breaks :

Long breaks mean , you have done your work and now it's some other agencies work to complete the balance work and while doing so , keep reporting to you about this progress !

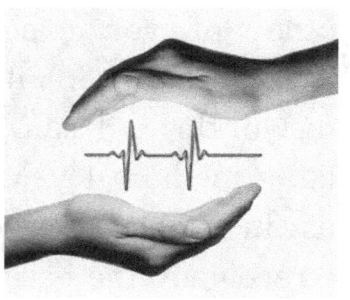

Courtesy: Geralt, Pixabay

Many time , researchers work with cross functional research fellows and institutions ! In recent pandemic , one may have seen , how the joint working in between various research institutes takes place ! One agency just receives the sample and after initial data filling transfer the sample to next research lab ! Here ,sample is tested and its test report is forwarded to another control center ! That control center analyses the report and suggest treatment plan code to first lab from where sample is received ! The control center apart from first lab , share the report with other statutory bodies and may be international institutes with which they have made formal agreement ! International institute , in turn record the status of particular phenomenon as compared to status of other nations and accordingly guide other help center to provide adequate resources to that particular country ! The help center in turn take the note of stock available with them and discuss for future requirement with another fulfillment unit ! The fulfillment unit contact the list of present manufacturers of that product and thus the required stock is supplied and consumed at local level by completing distribution cycle !

So, how much time goes in this process ? This is called as a long break ! Every researcher like to reduce this chain of communication but if the challenge and spread is universal , you need to wait till long time because such clearance decisions are taken very very cautiously keeping every institute's scientific role into mind ! If haphazardly incorrect decisions are taken then two things happens ! First thing happens is your decision become incorrect and second thing happens is that no one come forward to take the responsibility of incorrect decision because of any type of fears in their mind ! To avoid such type of issues , it's always better to use long break mechanism in place and let every decision maker take the critical strategic decision carefully after evaluation of risk of that decision for public life!

7.5 Research Gap :

Everyone knows that right from taking admission to research course to presenting your thesis is a time bound programme ! Researchers invest three to four years of their life to crack the code of scientific research !

But what happens if your research does not complete even after three to five years of time ? Can this happen ? Yes , this can happen in some of the toughest inventions ! For such researchers , the research institute need to extend their support and association noting the work done in last three to five years !

Have you ever wondered about the progress of various space missions ! In past years , with scientific zeal and policy makers ambition , scientist work with heart and soul to make these mission successful ! In first attempts , everything was risky and riskier , but when the final interaction is happened with actual live missions , people noted the typical changes in space and noting them they modified the plan for next space missions !

Courtesy: CFV, Pixabay

Till this modification , the launching of new spacecraft was kept at halt ! The modifications were happening in earth's manufacturing stations and when their earth trial become successful , then they launched it to planned space center !

So, how much mission gap was there? This can consist of several months to few years! This gap was essential to work most suitable designs, addition of more work packs, collecting more details, addition of manual connection, adding manned missions' essentials and many new things that will make new launch extensive and more inclusive!

In later missions, researchers also tried to increase the time at space center to study the space activities and happenings more clearly! When you observe things for longer duration, your understanding of those things become clearer and then your opinions become constructive! In fact, it's the special ability of researchers to wait till the final inference get generated in that experiment!

So, after noting the story of research breaks and gap, one can see the typical time management skills required in futuristic research and incidence-based research to become ready before the challenge become heavy!

In next step, we are going to discuss the story of logic! ✷✷✷

MULTIPLE CHOICE QUESTIONS

1) Identify the research break in the given examples !

A) Study is halted after 50% completion for 3 years !
B) Experiments are kept on hold for receiving output of other experiment for 1 day !
C) Trials are stopped for 1 year because of observed safety issues .
D) New production is stopped because of repeat system failure at peak loading condition since last 2 years !

2) Identify the example of gap in above example !

A) Research is on hold for three years because of lockdown !
B) Research is on hold for five years because of increased research funding budget !

C) Research is on hold for five years because of abnormal field activities are observed after project initiation !
D) All of the above

3) **Which of the following reason can become cause of research break ?**

A) Test report approval is pending from approving agency for 2 days !
B) Samples are collected but testing is on hold because of lack of resources since last 5 days !
C) Government approval is pending because of back-to-back 3 days holidays !
D) All of the above !

4) **Breaks –**

A) Refocus researchers
B) Refresh researchers
C) Reunite researchers
D) All of the above

STEP 8 : STORY OF LOGIC

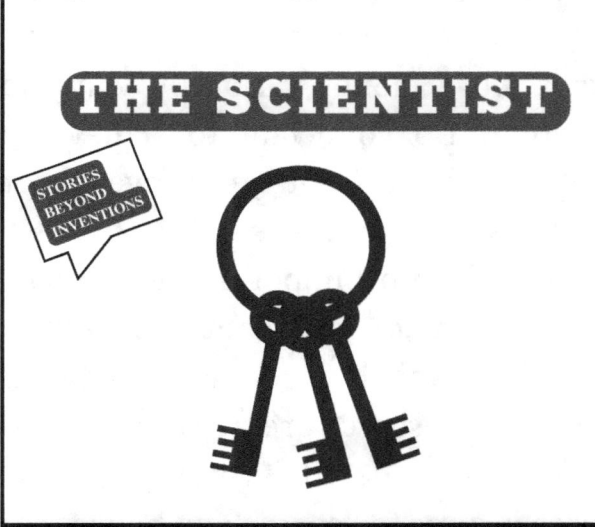

" So, which theory you will select first that will start your research work ? Which theory you will refer in between to cross check the progress of your experimentation ? Which theory you will attach as supporting literature before deriving your own research formula ?

Image Courtesy: Open Clip Art , Pixabay.com

8.1 Introduction :

Friends,

We are stepping into eighth step of this book and this step is about – research logic !

How do you decide to start research ? Do you want more money ? Do you want special recognition ? Do you want to achieve higher designation ? Or do you just have enormous interest in carrying out research ?

This is the most important decision for any researcher ! Why do you want to enter into the research field and how dedicated you are to complete your research are two things which are expressed by your detailed research publication ! Research is done for people and hence research logic is very much essential so that it can meet public expectations !

8.2 Initiation of Research Logic :

This is the first and fundamental step in any research process ! As we have said earlier, the research is a public work and hence it need to be thoughtful, useful and logical so that public can receive its intended advantages and can stay

Courtesy : KDBCMS, Pixabay

away from possible disadvantages ! To achieve this 'what is receivable to public ' and 'what is not receivable for public ', research logic is important !

Generally, people develop a typical field interest and they get admission to those courses ! As they keep studying, they get closer to useful applications of their course material ! This useful application fulfills public need and in return public pay the price of that application ! This exchange of knowledge enabled application with respect to self-earned money is the motivation behind every research work apart from inherent field interest !

If you don't have required field interest and you have entered into research field just to gain other type of recognition, then your research work will take hail lot of time, you will face miserable failures, your research guide may not co-operate with you after certain observations and you may need to leave the research midway in between ! So, development of field interest is very much essential before entering research field ! When you have field interest , it is the exact result of your inherent research logic of that field !

Someone having fond field interest in cricket or sports can develop suitable health products which can take care of serious injuries ,accidents and forced surgeries ! How the field of medicine and sport are related to each other is a well-known fact ! Every sportsman has to appear for his or her compulsory medical examination before playing any game ! The fitness level of sportsman are key performance indicators and if they are not fit for the game , their expected performance gets hampered ,which in turn affect team's performance and hence compulsory fitness test is must for every sportsman !

When this researcher has keen field interest in sports , he will learn the typical stressful positions and causes of various sport accidents that happen during a game ! It can be shoulder injury , head injury , fingure injury , ankle injury , chest injury , back injury , eye injury , chronic head pain , chronic muscle pain , situational anxiety and moments of discomfort before game ! When you are actually playing that sport , these feelings are experienced by you at this point of time or that point of time and hence when you actually decide to carry out research in this field , you find out solution for some of your sport pains and some of your friend's sport pain which equalizes to 'research in sport pain !' Unless ,you have field interest , your research logic will not orient towards right approach and hence your experimentation will not provide satisfactory results !

8.3 Logic for selection of theories :

When you are carrying out fundamental or applied research , you need to acquainted with various research theories established and published by other researchers !

So, which theory you will select first that will start your research work ? Which theory you will refer in between to cross check the progress of your experimentation ? Which theory you will attach as supporting literature before deriving your own research formula ?

These questions are some of the important logic questions for selection of research theories ! Sometimes , while carrying out research in particular field , you need to refer theories of various fields ,or more specifically you can say theories of allied fields !

Courtesy: OCA, Pixabay

If you are conducting research in thermodynamics , you need to know the theory of thermal and chemical equilibrium , you need to know the heat transfer principles and number of heat transfer formulae , you need to know material used for thermal applications , you need to understand theory behind thermal equipment design and insulation requirements , you need to understand theory of pressure distribution and breathing spaces for safe thermodynamic action

! This way you have to combine various theories before presenting your research work !

Additional knowledge of mathematical concepts like calculus , limits, integration , series , graphs , also help you to deal with practical formulation techniques ! This knowledge practically accelerates your research work and make it fairly presentable !

With this understanding , one can easily note the importance of stages of fundamental basic school level education ! If your most easy steps of school education are sound , you can easily enter the stage of primary education after pre -primary education ! When you complete primary education easily , you enter secondary education swiftly ! When you enter higher secondary education , you again feel like, attaining full of required level of knowledge ! When you complete diploma or degree level course with good marks which is result of your field interest , it reflects your association with chosen field is strong ! Once you enter into master's level education , may be after taking some experience , your knowledge spectrum again increases and you can readily relate to various concepts !

This is the stage of research which prepare your fundamental ground for last 30 odd years ! So , current education system , gives so much importance to human life that before completing these under graduation and post-graduation programmes , you cannot enter into research field ! This is the testimony of the fact that research is public domain work and you need to take utmost care of what you write after carrying out well oriented experiments ! You are recording universal truth and hence research work is always a notable work for mankind !

So , what will happen , if some bright genius of small age researches something extremely important to human society with his intelligence and now he want to present it to the world ! Can world of researchers will allow this kid to play on research field ?

Friends , this is the most liked talent in research field as we have earlier seen – the born intelligent kids ! They have exceptional intelligent quotient ! They cover up primary and secondary level education so fast by gaining knowledge and appearing for exam and thus they enter research field at very very tender age !

Later they complete their research work because of their intelligence and logical approach and thus serve this society for long time ! So , research requires intelligence and logic , age is just a number for research !

8.4 Logic beyond research failures :

How do you start your research after several failures ? This is nothing but the logic beyond research failure !

When you have multiple plans of dealing with research failures , you don't have to bother about the frequency of failures ! Ideally , there should not be any failure if you are logical enough before taking a particular research step ! But in reality , since the research work is done based on many assumptions and expectations , each time your assumptions and expectations may not turn into your favor and in all such situations , you face failures !

The logic beyond research failure is gather your courage and stand again to start research with new addition , new movement , different structure or may be out of the box thoughtful trial !

Be it a sport, research or normal life, if you are facing failures, they are result of incomplete study of the subject ! If your study is correct, your memory supports you during examinations and hence your trials become correct !

Nonetheless, if you are not ready to cover yourself and go ahead, it become a basic problem of attitude required to prove yourself in the research field !

Courtesy: Nikin, Pixabay

So, who is the best person, who checks whether your research logic is correct or not and thus protect you from unpleasant failures ? It's your research guide, who has spent whole of his or her lifetime for that field and hence they know that field totally !

In upcoming chapter, we are going to see the story of research intention in broader manner ! ✪✪✪

MULTIPLE CHOICE QUESTIONS

1) **Identify the most logical step to given research examples :**

   A) Theory – Aim of Experiment – Inference
   B) Field Study – Problem Analysis – Research
   C) Public Feedback – Sample Collection – Research
   D) All of the above !

2) **When a research experiment fails miserably ?**

   A) When research logic is incorrect in the very beginning !
   B) When research logic is partially incorrect !
   C) When research is started as per urgent need!
   D) When research receives low research funds for its completion !

3) Which theories are important before starting research in any field ?

A) Theories of material !
B) Theories of design elements & design calculations !
C) Theories of construction & rules of construction !
D) All of the above !

4) In which of the following situation research logic beyond failure is most important ?

A) A space mission is failed in its last few minutes due to crash landing !
B) A under construction bridge collapsed within a year of its construction and before building new bridge detailed research to be done by prominent technical institutes to find the root causes of this high budget failure within such early tenure !
C) When constant human deaths are seen because of rise of typical throat infection!
D) All of the above !

STEP 9 : STORY OF INTENTION

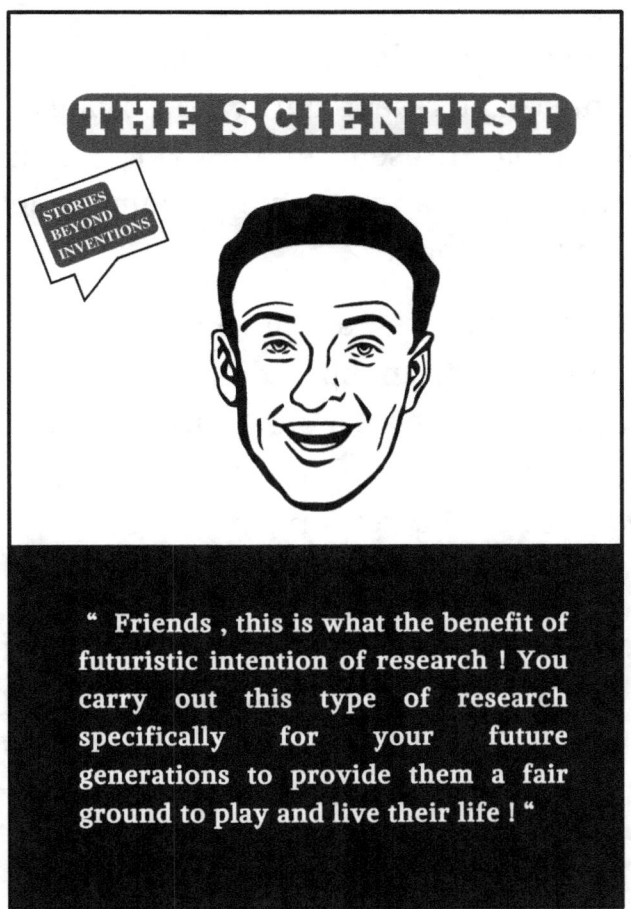

Image Courtesy: Open Clip Art , Pixabay.com

9.1 Introduction :

Friends ,

Research is purely a scientific activity which is carried out by some of the best intelligent minds present in this world ! Scientists and researchers work so well to invent new things present in this very nature that normal people can enjoy their benefits to unlimited extent !

However , application of researched inventions in public domain is restricted decision and the whole control of its issue , regulation and prohibition is taken by governing bodies of public care units !

So , when you are researching something new and innovative , you have to consult with your governing organizations about your plan of conducting that research and your sincere intention behind that invention !

In this step , we are going to discuss , how research intention mobilizes governing agencies support and mediation to ensure the research and its benefits reach to right people and the undesired effect of research are avoided to enter into public domain by all possible means !

9.2 Research for Positive Intentions :

Courtesy: Dee Mar , Pixabay

When research is carried out to derive positive results of that work such as creation of job opportunities in new field , provision of additional fuel for current range of fuels , addition of new farming technique that increases productivity and quality of crops , development of new software that reduces cycle time of traditional processing , all such type of research is known as research done for positive intentions !

Here , you inform about your research topics to institute where you are carrying out research ! The institute in turn share the research topic with research board under government's regulating capacity ! In addition , research institute also showcase research topic on their official websites for international referencing ! So , at any point of time , any visitor can easily see , what type of research is going on in respective nation ! The use of transparent communication also enhances mutual respect between two or more nations !

Suppose , one nation is carrying out research with positive intention , at very first time , your neighbor nation will keep watch on these activities by all possible means because he has always a concern about how you will use that research when it will reach its success stage ! Whether you will use it for your own growth or whether you will use it to threaten their existence ! Such matters are very very delicate and hence international referencing help such type of research to go ahead without any hindrance as such !

When , you complete this research , you publish relevant literature and world get understanding about it ! Fair research work that has international appeal and feel is felicitated in international knowledge platform and with this felicitation both the person doing that research and the nation which has observed that research

, both get acclamation for this noble and constructive work !

Research done with positive intention always serve people in good faith and spirit !

9.3 Research done for taking care of negative intentions :

Is this type of research really exists ? Yes , this type of research does exist and it is done to protect yourself and your land from attacks and harms which are done with negative intentions by your enemies and their allies !

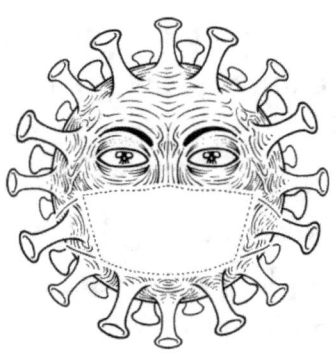

Courtesy: GDJ, Pixabay

In recent pandemic wave , people have experienced unprecedented effect of killer virus for which the vaccine was not developed before its infection ! If that vaccine was available before , chances of spreading infection and resultant death numbers must be reduced !

Many theories are discussed on international health and government platform, but tracing real origin, tracing spread of virus from one nation to another nation and its effect on innocent nation remained as an open book story that has no realistic end as such !

So, how such type of deadly virus can suddenly disturb or reorient the flow of nature ? Which perspective you will prefer as governing body of your nation ? Many people have seen that when there was strict lockdown, the pollution level reduced to drastic extent and the animals which were living in jungles started roaming in civil societies !

People have seen that, because of lockdown, people could give time to each other and for their relationship and this has significantly improved interpersonal understanding ! It is also seen that when more people are present in home, due to lack of freedom to move outside, people started feeling uncomfortable and it also highlighted the fact that when you are living in this surrounding, you have to preserve this nature, else if this nature goes beyond control of human being ,then its harmful impact will be out of control for everyone !

People said , this is the self-controlling mechanism of nature after every hundred years ! People also said nature balances huge increasing population by spread of such killer infections ! But one thing is for sure , futuristic research only has capacity to cure people from such spreads and hence role of research in curbing negative intentions become vital and for this all scientist and researcher's community has to learn the new methods of scientific research being done in different advanced countries to protect your country in case of adversity !

So , development of life saving drugs or vaccines is always research done to curb negative intentions ! Such type of research has enormous use for humanity ! Currently research is going on to find medicines on brain related diseases , heart related medicines , blood defect related medicines !

If this research is not done ,people will surely lose their lives because of spread of specific infection ! People have seen , when a contagious disease spread in a hugely populated country, the respective health system comes under tremendous performance pressure and when this maximum pressure level is passed then one cannot measure loss to people's lives and financial assets ! Its huge and unaffordable !

So, this type of research is directly supported by government ! Government allows team of prominent scientist to look on public challenges and provide reliable solution ! When the solution is derived, government provide it to people and thus people stay safe from such type of deadly spread !

9.4 Research done for futuristic intention :

Do you know your future ? Do you know your countries' future ? Do you know this planets future ?

If such type of questions were asked to any school or college going students before 20 years back, they may not answered it

Courtesy : DA , Pixabay

correctly but by looking at today's scientific development, one can forecast that in last space mission, scientist succeeded soft landing on moon, they also started study of sun's surrounding, so in upcoming future more research can be done in this subject of interest !

Before 20 years, there were only simple mobile phones, after 20 years from that time smart phones are almost become a mini computer and mini web market from where you can virtually buy and sell anything !

Before 20 years, travelling using air plane was very very costly, nowadays, many people are travelling with airplane !

So, what is happened in last 20 years because of which these things become accessible for public by paying its price in your own currency ?

Friends, this is what the benefit of futuristic intention of research ! You carry out this type of research specifically for your future generations to provide them a fair ground to play and live their life !

So, if someone now taking care of development of high yield crops , its benefit may not be received by current generation , but the upcoming generation will become its consumer ! By the time future generation become able to own advance things , your research will prove as a stepping stone for their overall life journey !

In the end , why the research is done ? To stay comfortable and safe against natural challenges which can happen any time ! If you are aware about natural changes in advance , you can protect yourself in advance , it is as simple as that ! And if you are not doing enough research for future needs , some other country will invent that new thing , will takes its international patent rights and thus set its supremacy on that invention ! For certain protected years , no country can use that formulation without their written consent and if someone try to do it , they will be diverted to applicable legal proceedings !

Hence , for any type of patented research work , other people have only one right to use that invention and that right is pay the money and buy those inventions for your use !

This is the formula of capitalism and bigger part of developed economies uses this principle of research and patenting to prove their supremacy on that typical subject matter !

So, if you want to really progress in highly competitive world , you have to increase your investment in research domain ! Only research power can fuel the spirit of entrepreneurship which can in turn produce more employment in your country and thus they can engage your own manpower with the help of which you carry out financial turnaround ! The financial development can develop your infrastructure and thus the overall growth momentum increases ! That is why , people always say , it's important to work but it's more important that where to work ! The work which gives you right salutation to your contribution is the best workplace indeed and field of research and development is one of the such fields ! So , as a researcher , always think with positive intentions but also take care of negative intentions to protect yourself and your nation ! Tie up with international collaborations and share the research which is going on to meet future demands ! Next step- Invention ! ✷✷✷

MULTIPLE CHOICE QUESTIONS

1) **Identify the research done with positive research intentions !**

   A) Technological new product development
   B) Health sciences research on new disease
   C) Environmental research on climate change
   D) All of the above !

2) **Identify the research done to take care of negative intentions of your enemies and their allies !**

   A) Research done in cybersecurity measures to avoid cyber hazards !
   B) Research done in futuristic journalism to avoid social divides !
   C) Research done in import -export trends to improve forex transactions !
   D) All of the above

STEP 10 : STORY OF INVENTION

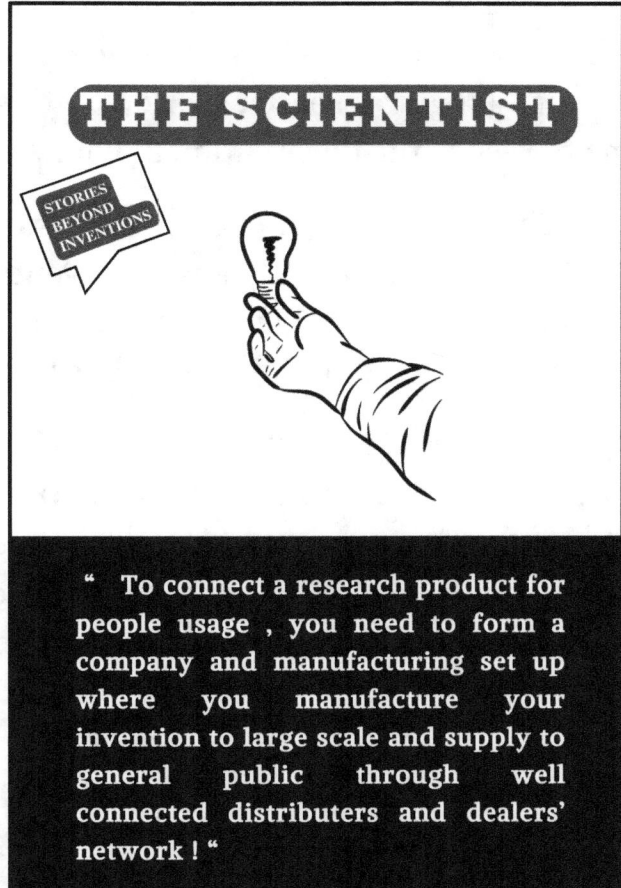

Image Courtesy: Open Clip Art , Pixabay.com

10.1 Introduction :

Friends,

Once we looked all aspect of research right from curiosity to intention, the final step deals with story of invention ! The ultimate result of your pro longed research work and this is the time to record the observation of your study !

The end result of every successful scientific research is called as an Invention and accordingly the terms describing that invention is theoretically and practically elaborated !

In every research, when theories are written, they are written with formation of key concepts. The major terms are defined and specified in words. The units of measurements are decided and they are prepared in line with international conventions for unit ! The apparatus designed for experiments are named after the invention or by the name of its creator ! Pictorial and graphical analysis is done to show the relevance of observed research to mathematical co-relation ! When research and mathematics joins hands in hand, the commercial viability of that invention become feasible ! This is because in commercial world measurement matters most ! Till you are not measuring , how can you sell it to target customers ?

So, when you invent unit of power as British Horse Power ,then only you can sell 1 BHP for 10 Rs ! When you invent unit of telecommunication signal in the form of memory capacity which is measured in GB ,then only you can sell 1 GB data for 19 Rs ! On the same line , when you research & develop the technology of oil exploration and drilling and when you follow unit of fuel measurement as Liter , then only you can sell 1 Liter of petrol for 104 Rs or one liter of diesel for 91 Rs !

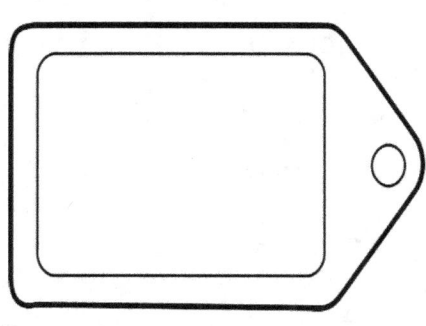

Courtesy : OCA , Pixabay

So , when you are selling some technical product , you are selling the comfort and ease behind that invention per unit value of its final sales price ! While deciding its sales price , you will take into account , the manufacturing expenses , raw material price , applicable taxes and expected profit for your research effects !

As the sales of these technical product goes on increasing , more investment is done in simplifying the product concept and adding additional features so that its future sales again go on increasing ! This is how one typical invention typically changes the size and shape of any ambitious organization !

To connect a research product for people usage you need to form a company and manufacturing set up where you manufacture your invention to large scale and supply to general public through well connected distributers and dealers' network !

Let us see ,different types of inventions as per their most popular categories !

1) **Scientific Inventions :**

These types of inventions are result of dedicated research done by individual scientist or team of researchers ! The prime aim of such inventions is to make living life easy and simple

! As per definition of machine, any arrangement that makes your work easy is known as machines ! Machine has a mechanism and that mechanism is based on theory of that machine and design of that machine according to theory written after natural and experimental observations behind the governing principle of that machine ! This governing principle mostly remain same on all part around the world and hence machine produced in one country cannot be sold to any other country without any hesitation !

Machines has a fixed service life and recommended guidelines of its fair uses ! Till the time, these conditions are adhered to, machine gives satisfactory results ! The moment deviation to guidelines is observed, defects in machine set up are observed !

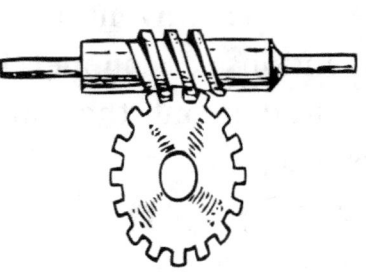

Courtesy: CFV, Pixabay

What type of defect a machine can automatically control is decided by research logic behind that machine ! In such research logic, you have to simply press the reset button to make things normal as per earlier set up !

Restoration logic help every user to retake the charge of machine in case of any deviations happened while its usage ! When machine do not start after reset or restart , there is internal connection mistake and for which people have to repair the machine to change the defective parts using available spare parts ! This is how scientific inventions are commercialized and made available regularly !

2) Defense related inventions :

These inventions can be scientific inventions , commercial inventions, artistic inventions but they are specifically used for defense needs to strengthen internal and external security !

Mass destruction weapons , new type of ammunition , advanced communication system and satellites , interpersonal communication codes , Earth – Water – Air defense transportation systems , defense financial software's , environment friendly personal protection and uniforms , such products come under defense related inventions !

Recently , a tent is invented for soldiers who are providing service in cold regions ! These tents keep the internal temperature warm and make that weather little bit comfortable !

Development of highly sensitive drone cameras for tracking man and material movement of boundary and mountain region can be considered as invention made for defense needs ! These inventions are made for nation and offered to nation as a symbol of patriotism !

3) **Health Sciences Inventions :**

Before scientific inventions made for technical product development , health and defense inventions comes to forefront ! First people need to live , then they need to be protected and then they can become productive to help respective nation grow in all possible directions !

Health science inventions typically deals with research on drugs, vaccines , treatment types , surgery techniques ! Surrounding wise , they deal with research on natural microorganism , bio-diversity , changes happening in natural eco-system and how the present species are struggling for their existence !

These inventions can be done for humans as well as for many animals ! Veterinary science has its own importance in taking care of animals !

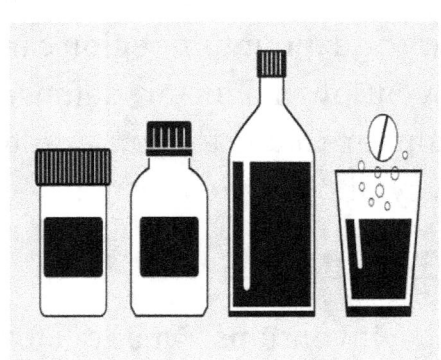

Courtesy: Segrest, Pixabay

As the health science predominantly deals with public service , the prices of these inventions are affordable to public on retail purchase !

However , scientist need to invest multi million dollars before a breakthrough health science invention could take shape in reality ! The increase in longevity of people , chances of fair life and welfare of society are some of the major objectives of every health science related invention !

4) **Outer world related inventions :**

Life on planet earth is becoming challenging day by day and now researchers are trying to find other planet where man can live safely and happily !

The need of variety of space mission is highlighting the importance of outer world inventions ! Currently , the space missions are carried out for dual purpose ! The first purpose is to keep constant watch on planet earth to protect it from outer world occurrence of natural phenomenon !

The second purpose is to explore the other planets and to find out the chances of human life there ! People are searching chances that , if there is risk to life on planet earth , they will move to another planet ! This is strong thought and success of few space missions in providing prediction that in near future , some breakthrough inventions can be achieved at least to make planet earth to other planet's temporary travel commercially feasible !

In the so called ' space race ' ,now achieving success of particular space mission is seen as symbol of scientific wisdom and indomitable willpower of internal research potential !

5) Commercial or Financial Inventions :

Have you seen the changes observed in carrying out commercial transactions since ages ? This is nothing but commercial inventions !

Starting from typical way of barter exchange to use of coins followed by currency notes in transacting items at the cost of money , now the stream of financial invention is using most sophisticated debit and credit cards for valuables transactions !

The recent inventions of dedicated mobile application with the help of which immediate mobile payments can be done is proving a path breaking invention ! It has lessened burden on physical currency and connected digital technology benefits to grassroot human life !

The uses of bit coin and crypto currency is also seen as new invention in the finance domain ! The use of internet and interlinking with forex trade is again changing the dimensions and scope of international business relation !

Earlier , if you have to exchange the foreign currency with your domestic currency , you need to follow a stringent paper-based process ! Currently with the help of internet , on opening a legitimate account with bank providing forex services , you can easily transact domestic national currency as per forex rate of that moment with your interested country of transaction ! This local to global connect is an important financial invention of last decade !

So , what can be future possibilities about global currency ? Can all nation come together and invent single global currency for dealing with each other ! If such type of invention will happen in future, then all the resources available with any nations will be transacted on single global currency and hence chances of financial divide will probably reduce to great extent !

This is so far about the story of inventions ! In next step , let's see , more about research guide ! ✳✳✳

MULTIPLE CHOICE QUESTIONS

1) Which of the following invention is most important to save human life ?

   A) Invention of vaccines
   B) Invention of critical surgery technology
   C) Inventions of drugs those avoid surgery
   D) All of the above !

2) Which of the following financial invention has most easy and most expansive retail transaction potential by using power of internet and mobile technology ?

   A) Digital Rupaya
   B) Phone Pay Mobile App
   C) Visa Master Card
   D) E- Gold

STEP 11 : GUIDE AS COACH

" Once the early fascination and other fancy ideas are settled after formal and informal introduction of this field , you will orient yourself towards knowing your research guide , their duty in your research scope , their expectation from you and your joint contribution to society by the time you complete your research work ! "

Image Courtesy: Clker Free Vector , Pixabay.com

11.1 Introduction :

Hello Friends ,

Welcome to the new section of this book ! This section will discuss the next 10 steps which are specifically related to the role of a research guide ! In first 10 steps , we have discussed, to best possible extent, which qualities are required for research student and how their research is expected to help society in which they are living ! In other words , in first 10 steps , we have done ground work of foundation required to enter into research fields through various available ways !

In this section , you have arrived where you were dreaming to arrive since long ! Once the early fascination and other fancy ideas are settled after formal and informal introduction of this field , you will orient yourself towards knowing your research guide , their duty in your research scope , their expectation from you and your joint contribution to society by the time you complete your research work !

This section will try to match the wavelength of researchers to their guides !

11.2 General Assumptions about research guide :

Every researcher when enters this field , he or she has some assumptions about their research guide based on their previous degree and post graduate project completion experience ! In the undergraduate and post graduate courses , they have carried out research which has comparatively lower application intensity !

11.2.A: If you think degree level research , basically you are given a research topic that will utilize your syllabus knowledge up to degree level subject ! These projects are basically related to modification of existing inventions to next level !

Courtesy: Tumisu, Pixabay

For example , if an undergraduate computer engineering student is conducting a degree or diploma level project work , it is specifically related to applications of computer science to humanity ! Application of knowledge of computer science that can resolve live problems ! How effectively you align the knowledge acquired in three years of diploma or four years of degree is tested by your new and thoughtful creativity ! You have to present a project which is not done earlier by anyone but specific to your own field !

So, the involvement of your project guide is very much limited in such type of project work and mostly as a degree or diploma student , you have to complete most of the project work taking major initiative ! Because , this project is your ticket to professional engineering ,technical or scientific world !

Your research guide may have academic experience of 10-15 years and industry tie up for 5-7 years ! Based on such combined association , they may direct you to relevant industry to observe how the practical work takes place and how your project can be done by using practical approaches ! The research guide will also provide access to internal college laboratory work where you can take trials of your project work before taking trial in industry !

This way preparing yourself in academic and practical environment , your diploma or degree level research guide open up your engineering talent for society in comparatively limited spectrum of growth !

However, some highly talented degree students carry out research of higher level and present their research to suitable forward-looking industry and secure their first job with approval of their project work from that industry ! This is how degree or diploma level project prove as first opportunity to show your research potential to the corporate world !

11.2.B: If you look the role of research guide for post graduate courses like MTech , M.S. , MBA , MD , they look at you not as a college student but a person with scientific expertise !

These research guides assume yourself as you possess fair knowledge up to degree level including your basic elementary fundamental knowledge ! They may ask you some questions which may sound like question of standard 4th but with this introduction , they will explore your post graduate level understanding and its different practical applications !

This is what research guides do at post graduate level ! You can converse them with basic concepts , they will not deny it , but with these questions they will ask you to go further beyond your basic curiosity !

Post graduate projects are basically field specialization projects ! Suppose you are a medicine student and you have completed your M.B.B.S. course ! While taking admission to post graduate courses like M.D. or M.S , you have to choose best interested specialization to heal the typical diseases observed in that field !

Courtesy: BHG, Pixabay

Suppose you choose specialization for heart or orthopedic , your two years post graduate studies will predominantly concentrate on these organs and their range of diseases ! From this study , you will come to know the basic functioning of these organs and what is the current status of patients suffering from diseases developing, because of infection or attack on these organs !

This type of post graduate study will always teach you how the normal organ functioning should be and how it is occurring right now , for that chosen specialization ! So, because of this specialized approach , you go deep and deeper till you understand the working completely and then your research experiments start to develop new and simple techniques which can treat patients at lesser operating cost , lesser hospital stay and quick rehabilitation to join their normal call of duty ! How successful a medicine's post graduate student become post this type of extensive research, determines his practical success, once he or she starts his own service hospital !

Once his or her post graduate specialization research is approved , he or she will become able possessor of that knowledge and with this knowledge , they will tell the world that they are expert in treating these specific diseases ! For initial period , they will take practical experience with some big hospitals as well as some rural clinics to understand the patient philosophy completely and when they will open their own center , they will practically take the charge of every decision which is beneficial for patient's health condition !

So , when you are carrying out post graduate research , your research guide as coach assumes you as if you are carrying out daily 2-3 heart surgeries if you are doing MS or you are prescribing medicines to 100-200 patient daily if you are going for MD ! That type of expertise is expected by your research guide and these expectations to be fulfilled by you by putting your total knowledge to given research subject !

11.2.C This stage of research and guides of this stage of research looks at you as a fellow serving the like-minded field ! Here knowledge equations are different ! You are carrying out doctorate level research and it has deep society connection for which you are investing your three to four years of life !

Do you read newspaper advertisements which tells the success story of research student acquiring doctorate in typical subjects ? Research of this level has huge society acclamation and knowledge attained at this level is most pure and proven experimental knowledge that make you extremely clever and alert about your surrounding !

So, if a diploma level candidate and a doctorate level candidate are interviewing for any multinational company's interview on same day , the designation offered to diploma level candidate and doctorate level candidate will have at least 10 steps difference ! To reach that designation , diploma level candidate has to spend at least 10-20 years' time in that field !

Courtesy: OCA, Pixabay

The candidate with doctorate level caliber will be interviewed for most futuristic department of the organization and he will work with best talent available in that organization ! Their daily interaction will happen with prominent customers , old friends of the organization , key statutory figures and exceptional academicians linked to organization since long long years !

This is the reason , the research guide as a coach consider yourself as their fellow who is supporting the field with more brilliant inventions and thus keeping its current identity relevant to society needs !

In this huge people's society , needs of the people keep changing and this focus shift is totally dependent upon the research work carried out in that field ! So, if you have to attract people towards your field and increase financial momentum of your field , you have to carry out path breaking inventions in that field !

If you look at trends of youth choosing particular field , it is clear that the selection is only because of future career opportunities and related financial freedom to earn for a decent living ! At the same time , youth also prefer to work in most safe and clean scientific fields where their health will be taken care of and they can enjoy the better work -life balance consistently !

So , in last decade , youth attracted to field of computers , IT and electronics and telecommunication ! With this influx of huge talent , look how the research in this field has reached pinnacle of its success ! You started with desktop and mainframe and now you have made your smartphone a mini computer with high memory capacity ! And if you see the financial freedom of this field , just note the top performing companies in global market , they are these tech giants ! So, this type of exponential but holistic growth is expected by your research guide as coach from you when you are carrying out doctorate level research !

11.2.D In the post doctorate level research , you have already acquired doctorate honor and now whatever you are doing is strictly brand-new invention ! This type of work may not be looked before anywhere or you have acquired so much knowledge from your study that now you can easily mix that knowledge with other fields to develop those fields ! Research guide at this stage looks at you as an inventor ! In next step , we are discussing ,Guide as friend ! ⊛⊛⊛

MULTIPLE-CHOICE QUESTIONS

1) What are typical expectations of your research guide when you carry out research at undergraduate level?

A) How you apply your basic technical knowledge to research subject.
B) How independently you carry out your research and how you report the progress.
C) How you take trial of your research subject in industrial framework.
D) All of the above

2) What is the most important thing for a post - doctorate level researcher?

A) Breakthrough Invention
B) Social Advantages of research
C) Cost benefit to society
D) All of the above

STEP 12 : GUIDE AS FRIEND

Image Courtesy: Handihow , Pixabay.com

12.1 Introduction :

Friends,

In last step, we have seen, how a research guide looks at you when you start your research work at various stages of graduate to post graduate research ! With increasing your age and knowledge level, expectations from you keep increasing and the gap between a typical guide-student formal as well as informal relationship become thinner ! A stage is reached in researchers' life when they start working with their guide on same subject of joint interest ! This is the level of confidence and caliber one guide expects from their research level students ! Research guide expects that students must acquire that level of knowledge from where the experience of research guide and knowledge of students make wonderful research for society !

So, in this step, we are going to see, how, when and why research guide become your friend !

12.2 How research guide become your friend :

As we know, becoming friend of each other means knowing each other , talking with each other , sharing like-minded interest , work towards common goal , be there in moments of happiness as well as testing times and in the end be respectful towards each other for what you both have and what you both don't have ! This is what the definition and meaning of friendship ! Friends ,accept each other as you are and they make your life easy till the time you are cherishing your friendship !

Courtesy : Pali Graphic ,Pixabay

In professional friendship, people come together for professional work ! They have qualification , they have talent and by coming together , they want to add value to that field !

When you enter the research field as a student , initially , your knowledge is tested by your research guide by applying different techniques which can be openly said or indirectly suggested ! In both of its form and spirit , you , as a student expected to find what is asked to explore & discover in that test drill !

It is the first impact on the mind of your research guide ! How easily and efficiently ,you produce the required task indicate your preparedness , simplicity , sincerity , dedication and commitment towards that work !

In next phase , when you and your research guide know each other for some time , you start asking some questions about the field development and you keep discussing about your own thoughts on such subjects !

This self-initiated interaction again makes your research guide happy and he start adding their input to your conversation ! Always remember , your research guide daily spends nearly one- or two-hours' time to note the latest update in your field as well as fields revolving around your field ! Such guide also has keen interest in basic general knowledge and hence they are always a well informed and updated person in the field !

So , even though the discussion is started from your side , by noting your research guides view on that subject you feel surprised ! Your research guide along with that subject's essential development ask you about future development in the field and how the society will receive benefits of this new development ! This is the identity of good research guide ! They are always updated about recent developments and they frankly share those developments with their students on formal or informal dialogue !

Your research guide become more comfortable with you when you present the practical research work in front of them ! This presentation is testimony of your expertise in handling modern scientific equipment's , noting their standard operational techniques , getting know how about their extreme features , providing information about the equipment uses in modern research ! This practical approach makes your research guide confident about your learning graphs and slowly they start co-operating with you just like a friend !

12.3 When your research guide become your friend ?

When you spend minimum 12 to 18 months of time in any type of research institute, you become familiar with the inherent methods of research and you become useful to that atmosphere ! Six months' time phase is considered as typical phase to settle in any professional environment !

In these six months initial phase, you have to carry out your admission and administration formalities within first 15-20 days ! In next 10-20 days, you have to adjust to your research

Courtesy: OCA, Pixabay

schedule and regular meetings with your research guide ! After this 40 -45 days' time, you have to start theoretical research on your subject and keep gathering important aspects before starting experimentation of your work !

In this phase, your research topic will be discussed by your guide and they will ask questions to get information about your intention behind that research ! If they notice the chosen topic is very very limited as far as current level of field advancement , they will either increase the study of same topic or they will suggest to add research on more allied topic along with chosen topic !

This is why the field of interest for research is very very vital ! If you are doing research in well set fields or may be partially saturated field , you will get extensive material for research theory but your practical experimentation will be limited , because many people has researched that branch of science and invented many new things after putting their efforts ! These inventions are commercialized and industry has also added their inputs to make that invention a popular one !

So , you have to choose a new research field ,where very few intelligent people are doing research ! Here your sharp intelligence will be tested and you will develop the theory of your research on your own ! Yes , the very clever inventor writes the precise theory of their research and afterwards people name that theory by their name as a mark of honor and prestige !

Your research guide become your friend , when you are writing such type of theory in your field of interest ! Since the research field is extremely new , your guide assists you in simplifying the scientific vocabulary about defining natural phenomenon , defining governing principle , putting forward the terminology and units of measurements of that inventions , benefits and disadvantages of that research and fields where that invention can find its application ! Ultimately ,all research is done to find out its applicability to current working fields !

When you conduct experiments to practically prove your theoretical conclusions , your research guide become your friend ! They see , how you set up the instrument , how you gather the required material , which codes and standard you research to choose right material for your research and then which processing is done to check the effect of addition of various reagents and reactants !

This overall preparation approach makes your research guide to think about your understanding of particular research work and if they notice some unusual things , they correct it as a friend of yours ! Once the experiment enters into critical observation phase , they join you along with other research guide to discuss the progress of your experiment ! Thus, during practical research work you can get guidance of more than one research guides ! This happens only because your research guide observe your commitment and he bring someone equally knowledgeable in the field to show them your talent ! This is nothing but start of entering into research network where because of your knowledge, zeal and commitment , you meet with some exceptional brilliant minds that actually help you to succeed in the research process ahead !

12.4 Why your research guide become your friend ?

As we have seen , once you spend time with your research institute for more than 12 months to 18 months' time , people start knowing you and now along with your intellectual capability , people also start looking at your interpersonal skills !

Your research guide can see, you are good as research student but best as human being! This is the reason why your research guide become your friend!

People who are blessed with excellent intelligent quotient are the people present on this planet as demigod! Because, like God, they can use their available knowledge for the betterment of society! They can solve people's issues by applying their research logic and thus they make things simple for people!

In this journey they also earn lot of fame and money but in the end, they get sensitive to note the social divide because of lack of education, proper nutrition and other conflicting issue of governance of that region and thus in the later part of their life when they are well settled in their field of expertise, they focus their attention on philanthropic work!

This broad spectrum towards looking at life attracts your research guide to become your friend! They notice this hidden potential in you quite early and thus always motivate you to contribute to higher level of excellence!

12.5 Friendly moments during research work :

As the research field is gathering of intelligent and brilliant top-class minds , there are many friendly moments that takes place during research process !

When typical research group is formed, they typically behave as a single integral unit with values of equality , compassion and freedom of expression ! Rather researchers are the only people who can listen to any theory proposed by other researchers and can raise a valid point if they notice some illogical hypothesis ! With their earlier knowledge , they will cross question the researcher who is presenting this theory and thus that discussion will later end to logical conclusion because of collaborated research team involvement ! Other than this , there can be perfect moments of joint celebrations on achievements of major or minor milestones , approval of particular theory , mathematical modelling of experimentation ! So , this is about how guide can become friend during research ! In next step , we will see ,Guide as philosopher ! ⊛⊛⊛

MULTIPLE-CHOICE QUESTIONS

1) As a keen researcher , how will you express your social inclination when you are developing bonds with your research guide ?

A) By sincerely completing research with social upliftment potential .
B) By researching solution to live social challenges .
C) By researching with social technological facilitation that will take care of huge population comfortably .
D) All of the above .

2) Research guide becomes your friend when

A) You take ownership of your research .
B) You stay updated about developments
C) You stay open about your research needs and fairly discuss at right time .
D) All of the above

STEP 13 : GUIDE AS PHILOSOPHER

Image Courtesy: Geralt , Pixabay.com

13.1 Introduction :

Friends ,

Welcome to the important step of every research work and that is the philosophical view any research possesses and how the research guide helps every researcher to understand the philosophy of what they are going to do ahead !

Philosophy is closely related to ethics and values that defines the scope and control of your research work ! Every researcher always needs to remember one simple fact that whatever they are going to invent in these three to four years is going to change the world than earlier !

The invention can prove as useful as well as harmful depending upon how the possessor of that invention utilize it ! And hence its duty of every researcher that when they complete respective research, they must set some permissions, some obligations and some controls to ensure the invention is just used for useful purpose to humanity as such !

This approach of discovering & using the invention for good is known as positive philosophy behind that invention !

13.2 Ethical Concerns :

Research is vast field and many ambitious leaders focus their vision on growth of scientific development in their country ! While achieving their aim & vision of steadfast growth , they discuss their plans with the core scientists' network , may be through well co-ordinated communication programme !

Courtesy : CFV, Pixabay

In such programmes, every scientist shares their ambitious future plans and express the need of policy and funding support from respective government ! Progressive government always know that if they invest 1 Rs in research and development activity , they are going to receive at least 100 Rs in next four to five years ,well co-ordinated research activities , across the nation !

This is because , when you invent something new and register for its patent rights , you achieve its exclusive rights for production and sales ! This sole production ability avoids any competition for that product and thus sale of that product become super easy ! If people like that product , they keep buying from you and thus one invention gives you global recognition!

So , the job of every progressive government is to engage their maximum manpower in nation building activities ! For nation building activities work is done from three -four sides , which are jobs created by government , jobs created by multinational private players , jobs created by retail service providers and jobs created by academic facilitation !

So , when a whole nation is working seamlessly on its research and development policy , after every year 1000 patents will get file and all inventors of those patents will start business of their own which will provide jobs to at least 100 people initially ! This is the ethical potential behind every research work ! Till the time , your research work is beneficial to society , government always support such type of research programme !

Research guide help you to find effect of your research on bio diversity present around ! If your invention is providing comfort to humanity at the cost of nature's degradation or natural habitat's lifecycle disturbance , then it's not ethical to carry on such research work !

13.3 Un -Ethical Concerns :

Power is the most desired obsession of its possessor ! Generally, either people are chosen for powerful roles or they grab that powerful role by any type of unethical means ! When such type of unethical leaders drives your nation , then the life of knowledgeable people become challenging ! This is because , the unethical leader always forces them to invent new things with which , he can threaten their enemy's and other people , so that they will follow his rule and thus surrender their freedom at his doorstep !

Unethical leaders are least bothered about effect of their decisions on normal human society ! They always think to dominate the

Courtesy : Lorem , Pixabay

region by using negative energy of their will !

Knowledge has many dimensions ! It is said that the steel knife which is used to cause harm to human can be used to treat & operate human being during surgery !

In the same way , you can have inventions of vaccines , drugs, medicines , technological aids and equipment's and on other hand you can have inventions of mass destruction weapons , pollution causing vehicles , food items containing harmful chemicals , misleading software's and things like that !

So, it is the choice of every government to impose stringent ban and prohibition of such unethical inventions because of which humanity and natural eco-system may possess severe threat of its existence !

What happens when such unethical inventions get released and it is seen many times ! Nature has its own response mechanism and if unethical inventions are released into environment , nature changes its normal parameters and try to balance itself !

The current challenges of climate changes , temperature rise , overflooding , air pollution was not there before 40 years ! Nature was nice at that time ! People were enjoying clean air , ample rain and descent breeze in summer time ! In last four decades , because of enormous tree cutting , greenhouse gases emission and changes in natural flow of water created issues of water clogging , excessive summer temperature and heavy rains that disturbs everything in just matter of few minutes ! This is nothing but nature's own response to changes already happened around it !

Hence every leader has to take care of the research work being done in and around his country ! If his neighbor country is carrying out research in some critical subjects , they have to use their intellectual caliber to check how the things are handled and if there are suspicious activities , they have to either take them during bilateral discussions or they have to use apt international forum to raise your valid concerns so that there can be enough voting on your side !

When world vote on your side , the unethical leaders have to face the agony of world power which include various types of prohibition and limitations ! This way the unethical leader loses their international credit and become the target for other nations ! Those countries limit their financial transactions and this ignites flame of disappointment in his own country ! When people's anger outbreaks , people change the leadership and thus new leader ensures such type of prohibitions are not faced by country in future tenure !

This is how your research guide can explain the philosophy behind any unethical research subjects and how researchers should never do such type of work !

13.4 Financial Philosophy :

Money is major motivation for many ! Many researchers put their all efforts behind research work to find path breaking invention ! When this invention is commercialized , money just keep pouring for their lifetime !

So ,how much money is too much money ? There is no limitation to this factor ! Money is just a number which can be measured in millions , billions and trillions and so on !

So , how comfortable you want to live your life decides the motivation behind your research ! This is the just one practical approach !

In other approach , your research guide convinces you to other philosophical aspects of your research work ! Your guide inspires you to do your research which will reduce peoples suffering in least cost , to invent something which can be bought by every common man , to invent something with which people can take treatment in less cost , to invent something for which people don't need to leave your nation and thus they can get what they want , in your nation only and that is also in most affordable cost !

If you look the growth of India's scientific development , India has invented few things related to space research in surprisingly lower costs and with same effectiveness compared to international benchmark !

India has shown to world that they can produce the cozy four-wheeler in a retail price of 1 Lakh ! India has shown to world that digital payment facility can be spread all around India and any person can use this facility if they have legitimate mobile connection and smartphone ! India has shown to world that they can take patent of most productive agricultural breeds and thus they can fulfill the need of food inside nation to great extent ! This is not simple philosophy ! This is mammoth social science ! Just imagine , feeding almost 140 crores people two times or three times , is a superb work indeed ! You can earn money by doing any type of work but to make that food available for normal living , farmers have to produce record crop production and for this they need healthy breeds which has good yield generation capacity ! This is why the revolution of crop production upliftment is called as green revolution ! In the same way white revolution to increase the milk collection is implemented !

Courtesy: Amritendu, Pixabay

Your research guide helps you to understand this national need apart from earning money ! National scientists may not earn in dollars and pounds but the job satisfaction they receive when they can see that their one invention can feed 140 crores people for their two times food need , this tells the huge value of their social contribution ! This is the most explained life philosophy by every research guide to every researcher !

13.5 Individual Philosophy :

This is nothing but what you would like to do or what you will avoid to do in your research work ! There can be two situations ! In first situation , your research guide will tell you pros and cons of every research and you have to simply obey their suggestion !

In second stand , he will provide you total freedom to choose what is good and what is bad for your research and then observe your decisions from remote distance !

So , as a researcher, you also need to possess some philosophical understanding and you have to refrain yourself from carrying out research in unethical or misleading direction ! Because in the end , it's your decision ! ✶✶✶

MULTIPLE-CHOICE QUESTIONS

1) On which front ethical aspect of research must withstand under test of time ?

   A) Environmental
   B) Social
   C) Biological
   D) All of the above

2) Given a chance to tranfer the patent right to multinational organization trying to sell their products in your country & outside using the patented research work , what will be your first demand before signing the deal ?

   A) For my country , product price will be according to break-even point .
   B) My royalty will be X % per sales !
   C) My profit percentage should be Y %
   D) You have to make 50 years deal !

STEP 14 : GUIDE AS A QUESTION BANK

Image Courtesy: Jeresta , Pixabay.com

14.1 Introduction :

Friends ,

As we are moving towards almost half part of this book , now it is essential to discuss few topics that really add more value to this discussion on role of research guide ! These topics will reveal the skill a research guide possess to get best out of any researcher as well as he stays there whenever research work is witnessing highlighting moment of research !

Let's see , how research guides interrogation skill finally ends up with a question bank for researcher !

14.2 Introductory Questions :

These types of questions are asked by any research guide during your initial introduction to each other ! Some type of personal information , some type of educational background , some type of family background , some type of career interest and academic performance and the reason of entering the research field are some of the common introductory questions !

Courtesy: CFV,Pixabay

Students are expected to give most honest answers to these questions as they reflect your personality trait and future research guidance methodology !

To understand the importance of fair introduction , let us consider example of two researchers introducing themselves to their research guide ! One has done masters last year and he is continuing his research immediately after getting master's qualification ! He doesn't have any professional experience and he belongs to a rich , well-educated family !

On other side , the second candidate has 10 years of practical work experience and he has completed his masters struggling with humble family background ! He has fond interest in music and when he is not doing research , he joins their music club which hosts musical shows in highly reputed arenas in the city !

Now, which student will get more positive introduction to research guide ? Off course, the second one, because he is experienced and he can handle the research expectations more easily since he has already witnessed work pressure levels for past decade ! Also, his hobby not only act as his stress reliever but it also gives him different social identity which in turn give rise to more great connections and hence this personality can be termed as outgoing and go-getter personality which is a basic must have in every research aspirant ! You need to be multidimensional, then only you can mix experience of different fields to add value to your project !

14.3 Research know-how questions :

If someone is preparing for any civil services exams, then it is expected that the student knows the syllabus of exam, examination schedule, paper pattern, preparation strategy, daily study schedule, mock interview preparation, provisional merit, interested fields of services, etc. ! Then only, when such type of student take admission to coaching class , then the guide gives him admission based on introductory questions about temperament required for this highly competitive examination !

If the coaching class guide observe , student is not capable to sit for studies , if he is not interested in verbal discussions and various mock interviews , if he is not serious about the goal of becoming civil servant , then they may suggest to look for some different class ! This is because , in such highly competitive class , there is very little scope for wastage of time ! People are sparing most productive time of their lifetime and hence all involved students need to bear equal temperament required for preparation of that exam and finally cracking the exam with great marks !

Courtesy: CFV, Pixabay

More students crack the exam, more civil servants' nation get from such coaching institute and in turn coaching center receives great recognition for their efforts to provide good quality civil servants !

Almost same thing happens with researchers also ! The research guides are keen to know how much research you have done before opting for that specialization ! From which sources you have collected information about admission process and how you strategized your research entry exam qualification plan !

Then they further question you about choice of your research topic and how much research is done in this field in recent years ! Suppose you are doing research in nano materials and nano technology , then you may be asked questions about basic material science , the crystal structure of metals , slip mechanism in metals after application of unidirectional stress , heat treatments of materials , cold working characteristics of the material , then uses of different types of microscopes which can cover scanning electron microscope , digital imaging microscope , then they may ask about start-ups which are providing services of nano material , what is the current investment done in such companies and what is the growth potential in next five years , which theory is referred to carry out research experiments and which institute is collaborated with that start up ?

When such all-round questions will be asked by your research guide then it gives them your preparedness for starting research !

If you don't give these answers easily , it naturally reflects that you need more efforts and accordingly research guide allows you more time to polish your fundamental knowledge by practically visiting these facilities and gain as much knowledge as possible ! Research guide will provide you address of that firm and also make introduction of yourself with the owner of that start up as per their industry -academia partnership programme , this way you will get latest practical exposure to your research topic and once you note the fundamental work being done in that start-up venture , you will carry on your experiment partially in your research institute and partially in collaborated start up !

14.4 Research Analysis and Computational Questions :

Doing work is one thing and proving that work in front of huge chunk of educated people is different thing ! In research experiments , you need to support your research finding by applicable mathematical theory !

If your subject doesn't involve mention of mathematics , then its fine , you need to explain that theory with supporting science behind that research !

Mathematical connotation is easily accepted by the people because there is certainty and repeatability of the results ! Other justification is always questioned by the people because of considerable scope for subjectivity !

So , research guide will ask about the formula of your experimentation , steps you carry out to get the desired research input , how you mix number of equations after analyzing the data of experiment , how clean your logic is in deciding particular inference , how many cross calculations you carry out to see that equation is correct and it doesn't change although you try with different data set !

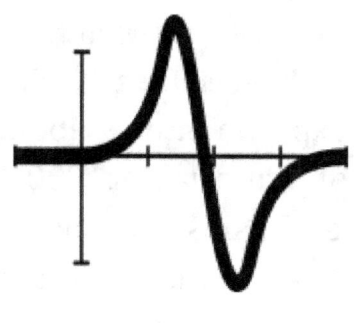

Courtesy: OCA,Pixabay

Data and graph have a close relation ! Every researcher is very much fascinated to see his experimental finding can be put on certain graphs ! Graphs are identity cards of your research success ! As security guard observe your ID card and allows you access to enter the classroom or research laboratory , in the same way , when you are able to put your experimental values on specific graph , it reflects not only mathematical connotation but visible spectrum of variations ! This is because graphs indicate variation in one or more variables with respect to variation in another major variable ! The graphical representation helps to check the interrelationship between variables , it derives the inference of direct or inverse relationship with variables and hence it helps to find out the final equation of the experiment based on graphical analysis ! Sometimes ,along with graphs , you have to also attach color photographs of your experimental finding as a full proof mechanism for computational analysis ! On such photographs , you write the status of experiment as per applied variables ! For example , if you are experimenting about microstructure changes of material at 500 , 1000, 1500 and 2000 degree Celsius , then along with graph of temperature variations versus microstructure phases , you need to take photographs of different microstructure observed with the help of microscope imaging techniques ! Because of this , you can easily co-relate the phase difference , addition of other elements during transition and the size variation in grain structure ! The observed mechanical value variation is nothing but result of these microstructure changes !

14.5 Presentation Questions :

These questions are about your final efforts to present the work done in last three to four years ! It can be about research report writing , screening and approval process ! It's about publication essentials and typical referencing protocol ! It's about what you have achieved in last three-four years !

In today's fast paced world , thousands of researchers are carrying out research in various fields and field people are noting this fresh talent through various platforming options !

In the age of e-commerce , once your research is approved and you have awarded doctorate , there are number of seed funding financial institutes who can fund your research venture ! All you have to do is present your research concept in front of interested venture capitalist and answer their techno commercial and other questions !

Along with practical applications , your research guide provides you details of all scientific journals and publications where you can publish your scholarly articles and where peer review and subsequent rating can be done !

Your research guide will also talk about how to avoid duplication in report and how to stay away from any sort of plagiarism in your report preparation ! In the end , the learning of three -four years need to be real and free form any type of doubts and unnecessary cross questions ! In next step , we will look into Guide as solo viewer ! ✹✹✹

MULTIPLE-CHOICE QUESTIONS

1) How often your research guide must ask you research know how related questions to test the status of your research ?

A) Daily Once !
B) Monthly Thrice !
C) Quarterly Twice !
D) Annually once !

2) If your experimental data doesn't form a systematic graph of inter-relationship of two or more variables , what you will do ?

A) Will repeat the experiment for error proofing
B) Will adjust values of observations
C) Will reduce scale of graph
D) Will reverse variables axis

STEP 15 : GUIDE AS SOLO VIEWER

Image Courtesy: Banger , Pixabay.com

15.1 Introduction :

Friends , welcome to next step of understanding your research guide ! In this discussion , we are going to see role of the research guide as Solo Viewer ! Independent observer of your research work ! Who has to represent himself as research guide of a prominent research institute that has tradition of certifying high potential researchers over the years and one who never stop learning till the final invention is discovered !

Every research institute has a reputation marking which is reflected through their number of inventions and their impact on society ! Do you think ,research carried out in global research institutes is different than one which is carried out in national research institutes ? Certainly No ! Everywhere in the world knowledge is same and eternal ! It cannot behave differently on different part of this earth ! If you add 2 into 3 ,its answer is going to be 5 in India also and in some other country also ! Because this is science ! This is pure mathematical science !

If we consider another example of scientific observation of water's boiling point , it's going to be 100 degree Celsius in India also and in other part of the world ! This is because ,the matter or material's intrinsic properties never changes ! They differentiate themselves from rest of the matter and hence these properties are known as their scientific identity ! So , different research institutes around the world ,observes- analyzes – cares natural phenomenon to such a deep and elaborated extent that their logic meets the intended invention !

So, in which type of research environment you are carrying research determines the research guides contribution to that field ! In this discussion , we are going to see , how high-level research thinking during research work possessed by research guide help to find best hidden talent from the candidate's research work ! This discussion will also spread light on how a researcher find out the most desired content in the presented research and how the modifications are suggested to make that report more promising !

15.2 Solo Viewing of student Caliber :

Research is highly intellectual work and research guides selected for such roles are extremely intelligent human being ! In school levels, 60 candidates take admission and 55-56 pass the SSC level ! In HSC again 30-40 students take admission to science side and 25-28 pass the HSC ! In graduation level course, 15-17 choose engineering while 10-11 choose medical side ! Before masters, 10-11 students immediately choose employment option and barely 5-6 students choose admission to masters ! So, when people complete masters and take admission to doctorate or PHD courses 2 or 3 students continue research, rest 3 students receive various job offers !

So, just observe what type of research caliber is available for research ! This is highly challenging field to enter as career and many people choose to settle in their life with

Courtesy: OCA, Pixabay

graduation or post-graduation qualification ! Their aim in life is to contribute to active field where products are prepared and sold and money is earned ! Initially ,this sounds good ,but after an experience of 10-15 years , you critically feel that you have to upgrade your knowledge base if you have to excel in your career ahead ! Because such positions required the most updated research knowledge with which organization can enter into new venture and thus, they can expand their business !

Your research guide looks at you as solo entrepreneur who can design product based on his research work , can decide its manufacturing sequence , can test the products as per stringent performance requirement , can install final product on customer site and then keep its maintenance schedule under regular check so that his invention keep working seamlessly ! This is what the exact requirement of any scientific research !

So , if you are entering with just graduate level knowledge , the knowledge gap between research aptitude and research scope create delay in your research ! Human memory has limitations and people tend to forget many things ! When you are studying till research level , you may forget many things but with constant touch with research guide you may remember forgotten concept and later you can co-relate it with your conceptual interpretation !

15.3 Solo Viewing of Research Scope :

When student's caliber is known the research guide challenges him with huge amount of research scope ! This research scope may involve understanding different theories , carrying out design and construction of experimental models , taking care of these models during initial laboratory set up , when the model is fairly set up in the research lab , then planning the subsequent trials and noting the observations ! Putting those observations into graphical models and check how it looks !

If some trials are not giving required result , then adopting to different approach with which some hopes can be seen ! Research guide always

Courtesy : CFV, Pixabay

analyses your decision-making capacity ! How you continue your struggle to fetch something new is always an exciting thing to watch for !

If the research scope of field of computers is considered , the earlier research scope was revolving around development of solid-state electronics ! The main aim of these studies was to observe the movement of electrons in particular direction and once its known then divert it to desired direction ! So , a well guided stream of electrons in the form of electric current is used to send electric signals to various components kept in electronic circuit and this in turn shows results on pre designed display screen !

Starting with smaller size and moving towards higher sizes , researchers has seen that development logic is repeatable ! As far as only display is concerned , people have seen cathode ray tube display , liquid crystal display , light emitting diode display and now OLED display !

The experience of viewing is continuously improving and so as the prices of product ! This is how research scope varies with the help of available students' caliber for research !

Once the solid-state devices are discovered , later software programming languages are evolved ! Earlier you started with Morse code , DOS commands , followed by programming languages like C , C++, Java ! Then with revolution of mobile technology , new operating system known as Android is evolved !

Addition of features is again researched and the mobile which was just used for communication is now being used for video recording , for using internet services , for calculations , for organizing and tracing activities , for using mobile applications , for using social media and personal messaging !

Each of this service is in fact is an invention carried out by electronics and computer engineer ! They invented these software's , taken their copyrights and sold the application to various customers ! A simple application like e-mail service is subscribed by trillion subscribers ! Why ? Because it has provided fast paced and cost-effective approach to traditional written communication ! You can write on word files and send them to intended recipient ! Worldwide web and hyper texting mailing protocol has made this communication channel possible ! The research done in networking of computers proved as backbone of this service !

So , how these services are traded ! Based on one time purchase cost of software CD' s and download link or by monthly and annual subscription plans ! The content was legally protected and hence in case of any copyright infringements , lawful actions were taken in case-to-case basis !

So , if you look back just for 20 years , these services were actually researched inside the laboratories , their trials were arranged first on known data , then on known-unknown data and once it become successful , it is released for public access !

Meanwhile, this type of research also experienced research challenges in the form of bugs, viruses and system errors ! Which are tackled by applying skill of software testing, antivirus programme creation and system version modifications ! All this work is done by researchers in their laboratories and final product is given to customer ! So, a research guide give you research scope in such a way that you study all relevant points and make a product out of it which can serve for people's needs ! In case, if some issues are noted in product, your research must find a legitimate solution ! That type of research logic is desirable !

Courtesy: Jcamargo, Pixabay

15.4 Solo viewing for promotion in research :

When in initial first two year, you show satisfactory research aptitude ,your research guide promotes you to higher level of research !

In fact, when you get grip of your work, you initiate many creative ideas and just try those things to discover how it moves ahead ! This experimentation habit is main crucks of the matter and every researcher possess it to different extent !

That's why, during your research work, you have to keep two things always handy, which is your imagination and next is your creativity ! Imagination helps you to take a lookout of various experimental possibilities while creativity shows you how things can be assembled ! Unless you imagine, it cannot be created and unless you create it, it cannot be trialed and observed ! Unless you observe, you can not equate and formulate and unless you formulate, it cannot be proven in laboratories ! Unless lab trials are successful, field trials cannot be taken and unless field trials become successful, its commercial production cannot be started ! In this journey, your research guide ,help you to explore all possibilities and try relentlessly till you achieve the desired research goal ! In next step ,we are going to see, research guide as reliever ! ✪✪✪

MULTIPLE-CHOICE QUESTIONS

1) As a solo viewer of your research student ,which point you will specifically check to ensure research is going right way ?

A) Practical Implementation of research
B) Detailed theoretical coverage
C) Scope for future upgrades
D) All of the above

2) As a research student , on which aspect you will focus your total attention so that your research guide views your report with ease and appreciate your efforts ?

A) Research Intention clarity
B) Conversion of numerical data into easily relatable graphical presentation
C) Error free formulations
D) All of the above

STEP 16 : GUIDE AS A RELIEVER

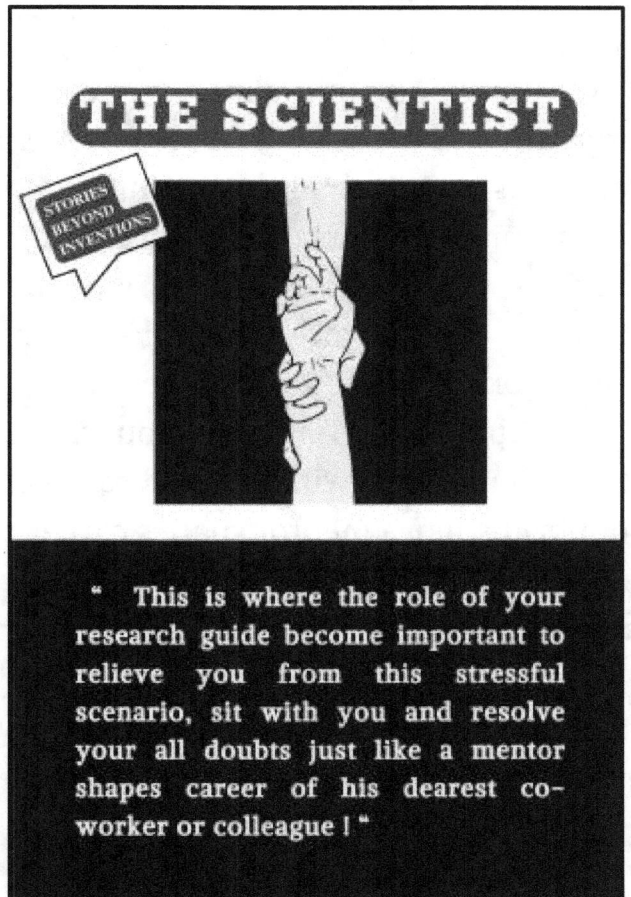

Image Courtesy: Milaoktasafitri , Pixabay.com

16.1 Introduction :

Friends ,

Research is challenging field and here lots of testing situations arises as your research moves ahead ! When you are constantly facing hurdles in your research , you get tired and worried about your research work completion ! You are so much closer to your invention but final steps are not getting realized ! This testing moment is quite stressful to new researchers who are just entered into the field after doing their masters ! This is where the role of your research guide become important to relieve you from this stressful scenario, sit with you and resolve your all doubts just like a mentor shapes career of his dearest co-worker or colleague ! In this discussion , we are going to see some of the basic testing moment during research and how your research guide makes things easy for you with their right mediation and resolution of issues !

16.2 Relieving during literature search :

As a first step of realizing your research work, you have to carry out extensive literature study ! Without this, you cannot decide aim of your research ! Even though you are coming up with new concept, you have to find out its basic supporting concept, then only you can co-relate with different terminology during research phases !

Courtesy: OCA,Pixabay

If you consider the origin of mechanical engineering, it was basically related to uses of material to make machines by carrying out their construction with many elements arranged in systematic manner and driven with one or more driving mechanisms which cause respective research logic being delivered to show final effect of that machine !

So, they first invented wheel, when wheel is further researched, then invented single wheel cycle, after that two-wheeler cycle is invented , which is followed by development of bullock cart from where two-wheeler is upgraded to four-wheeler ! Later on, development of IC engine and EC engine given rise to technology of automobile by using chemical energy of petroleum product to drive vehicle by mechanical energy ! And now ,we are living in the age of Formula 1 racing cars and sport bikes which are driven at a speed of 180-200 Km/Hr. !

This is what the research is all about ! In the initial phase you struggle with designing basic mechanism ! Lots of theories need to studied ! Lots of trial calculations need to experiment . You have to understand the forces acting on machines and conditions in which machines are expected to function ! You have to identify which is the critical mechanism , which is supporting mechanism , which is conversion unit and which is control unit ! Every machine has such type of mechanisms to support the final function of that machine !

So , when you carry out literature study of electric engineering required for that machine development , metallurgical study required for selection of right material , machine design and stress analysis theory to find out its maximum design limits , sound engineering concepts to keep the noise level within control , when all such theories are thoroughly studied, then only you can use scientific principles precisely by which your machine logic will get develop !

Your research guide gives you timely hints of relevant and supportive theories which you need to read and go through to crack the research logic of your experiment ! In theories , previous researchers have done some work and they have observed some peculiar phenomenon of experiment under consideration ! So, when you read that literature , you get your homework ready to face that peculiar situation !

In some machines , if some opposing forces are present , then you also need to read that literature so that you can protect your machine from such opposite or reverse research logic !

In aerodynamics , your helicopter or air plane is supposed to lift upward against the gravitational force ! If the inclination to ground during flight is not correct , the required altitude is not achieved and which can cause launching issues during take off ! In opposite case , when the flight is landing to its base station , the speed and altitude variation must happen systematically else there is always risk of uncontrolled landing which may experience jerks inside the plane !

So , as a researcher , you have to take all factors into account to create a full proof machine that meet your research goal ! Research guide tell you list of subjects which you need to master before designing your advance machine !

The research has two dimensions ! Initially you invent a new product and then goes on upgrading with extensive research to produce different models of that invention !

In other format , same researcher goes on inventing different machines in same field and later their team develop individual invention ! If you typically look at electrical inventions , there are number of inventions researched by same researcher ! You have electric fan , electric iron , power transmission mechanism , generators, transformers , inductors , electric motors , electric switches , electric control panel ! The list is enormous !

When you move to electronics and telecommunication , you see inventions like diode, triode, resistance , capacitance , integrated circuits , voice and frequency modulation techniques , telephones , wireless communication , internet !

When same focus is moved to IT and computer , you can list out personal computers , laptops , tablet , digital sounds & headsets , memory storage devices , internal memory and external memory enhancement , hard disc , DVD, CD !

If same literature study is done for industrial automation , mechatronics and robotics , then again your focus will shift towards theory of programmable logic , NC & CNC machines theory , mechanism of robotics arms and their electronics at chip or microchip level !

A research guide gives these glimpses of subjects which you need to thoroughly understand so that when you carry out your experiments, he will just observe the approach with which you are continuing it!

16.3 Relieving during practical work :

How many times you have experienced that, you have carried out an experiment and you are stuck at some point! You just call to your research guide, to just see your arrangement and guide you about, why the results are not coming, the research guide simply look at it and just shakes a little bit and the result start coming ! Has this happened to you earlier ?

In another example, you have designed an internal circuit, fitted components and you are now going to take trial of that circuit by starting current

Courtesy : OCA, Pixabay

through it ! You start the current and you observe, within two minutes time, that circuit stops functioning ! You again take second trial for rechecking the things and you again observed almost same results ! When you call your research guide, they just see your circuit design and relocate two parts at their desired location ! In third trial, you observe, the circuit is working normally and it is working till five to seven hours without fail ! Its expected running time is four hours, the circuit is working fine for five to seven hours ! This is called as relief given during practical trials !

Not only to engineering and technology, but in health science also, your research guide helps you during crunch moments of your research ! If you are developing a drug and if the animal on which you are taking trials is showing some side effects, your research guide adjusts the current composition level to reduce such side effects and again when you take drug trial, the side effects get vanished !

Your research guide is a person with both theoretical and practical experience ! This is the reason when they notice your problem , the relevant theory strikes their mind and with the help of earlier research experience , they just apply their mind to make things working as per desired performance level !

It's always a moment of pride when a researcher carry out his theoretical and experimental work on his own and his research guide just look the final results ! It's the best proven potential of any aspiring researcher ! Such type of talent is highly desired by top notch industries !

16.4 Relieving during Publication work :

There are two aspects of every research work ! First one is you have done excellent research and full proof trials but your research paper is lacking some major technical justifications on which many peers asking questions and thus doubting your skillful work !

In another aspect , you have completed your extensive research , supported same with proven trials and repeat results , you have also prepared a detailed research report with the help of your research guide and it is published in leading scientific research journals ! In second aspect ,instead of queries from peer group review , there are noteworthy praise and acclamation which explain the details of how effectively the research work is documented and published with full linking to references , literature study , acknowledgements and special mentions !

Courtesy : CFV, Pixabay

Every research guide suggests their valuable research input to the paper which is published by the researcher ! There is need of specific phrasing which new researcher may not be aware but research guide exactly knows the chronology of reporting , presentation of data and pictures , experimental details and cross-referencing results ! This way , research guide tries to provide relief to student by adding all important points on which questions can be asked by other researchers and answer of those questions will be found inside that paper only ! In next step , we are going to see ,research guide as a simplifier ! ✸✸✸

MULTIPLE-CHOICE QUESTIONS

1) **By which means , your research guide will provide you relief during initial phases of your research ?**

   A) By suggesting important literature
   B) By suggesting supporting literature
   C) By suggesting conversion theories
   D) All of the above

2) **When you are stuck in the practical work of your research work , how your research guide provides you the necessary relief ?**

   A) Using their vast experience , he simplifies what went wrong things
   B) They use their research logic developed over the years to check and balance with all possible trials
   C) They denote few major miss outs in experimental procedures
   D) All of the above

STEP 17 : GUIDE AS A SIMPLIFIER

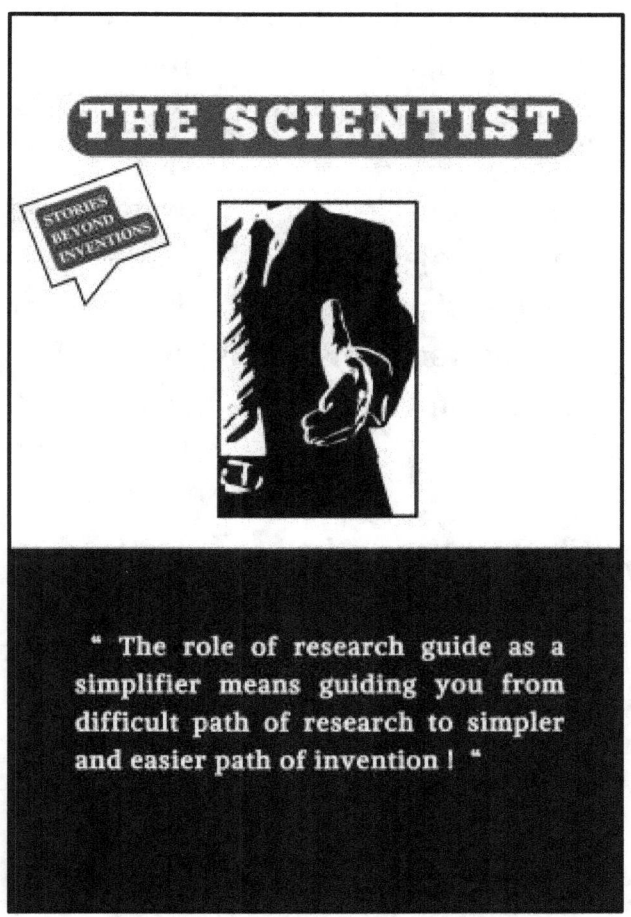

Image Courtesy: Vika Glitter , Pixabay.com

17.1 Introduction :

Friends ,

When you get a numerical sum of standard 10 or class of engineering's second year ,which one is seen a simple one and why ?

When you asked to draw a three-dimensional projection view of spherical part and other part having rectangular-circular and triangular intersections joined to each other , which projection you will prefer to draw & why ?

When you are asked to write few lines on your favorite hobby and on the current state of currency depreciation in foreign exchange market , which subject you will prefer to write and why ?

Friends ,

This is about simplicity and difficulty ! Simple things are easily achievable while difficult things need some sort of prior preparation in the form of study or exercise done with goal-oriented strategy and action plan ! The role of research guide as a simplifier means guiding you from difficult path of research to simpler and easier path of invention ! In this discussion , we are going to spread some light on simplification of difficult research tasks in today's scientific environment !

17.2 Simplification of Literature study :

Courtesy: Paul, Pixabay

There was a time in research history ,when getting research literature was very very difficult task ! First , every nation was not at same level of scientific thinking ! Some nations were struggling with age-old civilization concepts , people have different behavior habit , people were treating each other differently and in all these civil issues , scientific development was happening in bits and pieces !

As the scientists started finding success through their invention , new machines were introduced for world and people started noting the uses of machines compared to human efforts ! This, in turn, given rise to various versions of industrial revolution and rest is the history for mankind ! The rise of capitalism focused their efforts on sound scientific research and with this all inventions , products are created and sold throughout the world ! Financial exchange is happened and producing countries become rich as they improved their sales proportion to other countries where such parts are not produced because of lack of resources and lack of knowledge !

This is the reason that technical knowledge was fully protected by the companies producing those parts and till the moment of emergence of engineering colleges and technological institutes , this knowledge was intellectual property of inventors and their firms ! In today's time also , lots of knowledge is not made public and many organizations are utilizing that knowledge to create profitable products ! This is the whole gimmick of business world to earn money using technical knowledge

! With starting of engineering colleges , need of books was felt and protected technical research knowledge become a public access property through low to medium cost books ! The books providing special knowledge were sold at higher rate also ! This is followed by research paper publication press where monthly collected articles are printed and sold to various interested bodies ! With the emergence of computers and internet , electronic publication option become feasible and now whatever knowledge is available with scholars is become a public access property worldwide ! In today's smart phone age , you are just one search away to get the knowledge you require with the help of just one click !

    Your research guide as a simplifier for this type of research literature gives you unlimited access to their institute's library where traditional and advanced literature is available ! They give you access to institutes personal computer and high-speed internet connection with which you can search papers on specific subjects ! They can also made arrangement to provide public wi-fi services in the premises of institutes from where, you can access required website ! In some research institutes , there is friendly tie up with foreign research institutes ! Your research guide involves you in such groups where you can get the required knowledge and add your research knowledge also !

In today's video streaming age instead of reading 100 pages of a research paper , you can directly see well recorded interview of researchers taken by field experts where all the progress of their research is directly recorded ! This 30-minute video become base for your four years preparation !

There can be many education websites where you can join for number of skill courses and get certified online ! Such education website also proves as a source of research literature and advance courses ! What rest need to be done for more simplification ? Your research guide is always in touch with various social happening and they try to bring positive changes inside your premises for quick access to their students for their research work !

In the end two research guides are differentiated by their attitudes and actions ! One research guide follows up with their leadership to sanction various research budgets and other rules of the institute ! While other research guide has absolutely all freedom to bring most advanced things for students and budget or permissions are just formalities ! Until students are presenting their papers in international forum and maintaining the reputation of institute in international scientific forum , leadership is going to support and stand behind such research guide very very firmly ! Such type of research guides is best example of simplifier ! Isn't it ?

17.3 Simplification of Experimental Facilities :

The main difference in between a simple research and advanced research is the facilities provided to every research in their institutes ! What does mean by advanced research ! It is the next stage research where very few researchers are reached ! This early reach at required

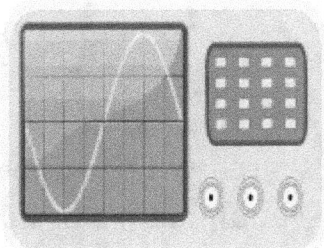

Courtesy: CFV , Pixabay

knowledge platform gives them early start to start production of research product and again sell instantly to public at large !

So , although a research institute who may be investing millions of dollars in procuring advanced machines and give its researchers total freedom to make most of it , that institute get the research breakthrough in quickest time and thus enters into commercial space quickly and gain the business momentum swiftly ! Secondly , people never forget the value addition of such firms because of their early and accurate arrival ! People stay loyal to their product and this in turn gives them great competitive advantage with respect to other players !

Whereas in a simple research firm , although any scholar may take admission , his research will take more time to finish because the equipment's or machines available there are not so much advanced or sophisticated ! Secondly , if some machine repairs occur , the institutes policy is not flexible and students has to wait for long time to pass through regulatory procedure ! In this time , other institutes student

,achieves important milestones and present their work to world !

So , as a simplifier , your research guide constantly give appointment to various scientific equipment manufacturers and receives the quotations ! He regularly visits various industrial exhibitions where advanced scientific machines are displayed . He gets information about such machines and brings some of them to your institutes ! He takes part in various industrial and academic conferences where current status of research is discussed and multiple solutions are presented in front of researchers ! He always tries to network with the people who has fond interest in scientific development and he also introduce brightest students of his institute to prominent and influential leaders !

Because of excellent connect of your research guide with fellow research institutes , he allows you for multiple training programmes in such allied research institutes which will add required knowledge in your research work !

In certain prestigious research institutes , various top-notch businessman and academicians present their advanced research work and this benefits the students attending the sessions ! The purpose of such sharing is to give instant guidance to researchers about which equipment's are currently preferred by industry and academic masters noting its accurate result producing capability !

17.4 Simplification for excellent career opportunities :

Every prominent research institute has strong tie up with various multibillion organizations which are always in search of research talent ! They can pay them par excellence package, facilities, perks and benefits ,just to get that candidate into their organization ! Such organizations are best examples of investors in human capital !

Courtesy: CFV, Pixabay

In addition to this , institutes may allow to carry research in line with few selected organizations with their experienced teams, this gives candidates early exposure to practical research and development unit and hence in future, not only in this organization but for any organization, they can continue their research !

If academic interests are concerned instead of business interest, your research guide may recommend your name to another prominent research institute as a professor or research scientist to share your knowledge to post graduate or graduate level students ! Once you receive doctorate, you can join and serve many technical colleges to impart the advance knowledge to future students !

When student receives guidance from highly qualified and industrially experienced professors, their grasping skills become easy and they can contribute to significant level because of clarity of that particular subject expertise !

In next step, we are going to see research guide as sophisticator ! ❋❋❋

MULTIPLE-CHOICE QUESTIONS

1) Which of the following task can be considered as simplification for your research work by your researcher ?

   A) Provision of conducive environment
   B) Provision of ultramodern equipment's
   C) 24 x7 access to research center with basic formalities
   D) All of the above .

2) In a simple research institute , the time required to complete the research is more in comparison to time required in advanced research institute because of -

   A) Lengthy internal approval processes
   B) Age old laboratory equipment
   C) Low motivation level for researchers
   D) All of the above

STEP 18 : GUIDE AS A SOPHISTICATOR

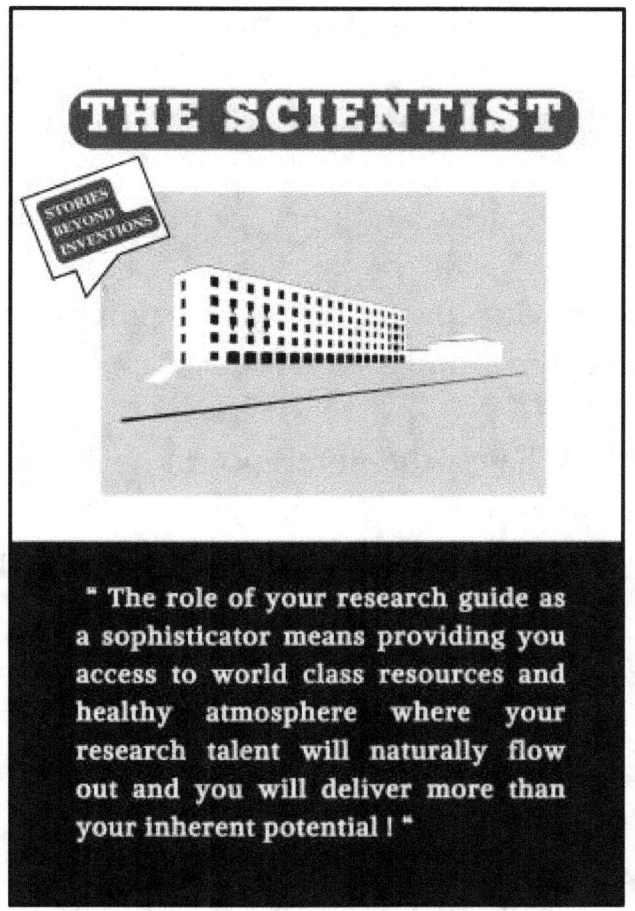

Image Courtesy: Open Clip Art , Pixabay.com

18.1 Introduction :

Friends ,

What is the difference in taste of a cup of tea taken in roadside tea stall , taken in three-star hotel and taken on the rooftop of a five-star hotel ?

What is the difference in quality of fabric of a men's shirt purchased from roadside stall , from a city's retail outlet and purchased from one of the brand's showrooms ?

What is the difference in between health treatment received form general clinic , society hospital and corporate multispecialty hospital ?

If you could relate, these three examples to your overall life experiences , you will easily understand the meaning of sophistication !

Sophistication refers to the way you look at every opportunity to serve , prepare the most appealing arrangement to deliver those services and in the end give your customer unforgettable or memorable ,extremely pleasant experience, with the help of utilization of high-quality resources !

The role of your research guide as a sophisticator means providing you access to world class resources and healthy atmosphere where your research talent will naturally flow out and you will deliver more than your inherent potential ! Because , what a pleasant atmosphere provides to a normal student cannot be provided by a poor atmosphere for same student !

On the other hand , if a scholar is working in poor environment for his research , he will possibly achieve half than what he can achieve in more sophisticated environment ! This is because , keeping the rate of his efforts same , there is addition to his effort from the environment in which he is conducting his research !

Every education and human resource development board tries to depute research guides and research in charge who have handled descent level of sophistication in past !

18.2 Sophistication inside research facility :

Do you know , some people just hate dirtiness around their working places ! They just want everything clean and clear ! Why this personality trait is observed at more high-profile positions ?

Courtesy: Ezequiel, Pixabay

Friends , this is nothing but the high level of practice of sophistication desired at such distinctive roles ! These roles demand different type of mannerism ! If you just want to explain with some simple examples , you can consider ,example of washing your hand in wash basin ! There is typical sophisticated method for this !

You have to gently approach the basin , you have to softly press the liquid hand cleaner push button , taking one or two drop of hand cleaner you have to rub it on your both hands , then you have to slowly start the running water tap, you have to clean both hands, you have to close the tap and hold your both hands in front of dryer ! If there is need to rub hands with handkerchief, you may do so ! When you carry out all these steps for cleaning your hands, its most sophisticated process of doing that work !

So, the facility which is providing you this service has made arrangement of descent wash basin with ample water supply, trending hand cleaner which is observed in many such high-profile places, arrangement of dryer which is again quite a comfort giving task, when all such detailed things are arranged for seemingly simple looking task, your system is supposed to be as sophisticated !

Taking forward the example of hand dryer to sophistication of laboratory equipment's, we can consider the process with which a research lab is maintained to make it highly sophisticated !

The instruments have specific place and they are kept at that place only after the work is finished ! The electric connection line is concealed and it has ample switches so that candidate can use any port for its connection requirement ! Computers are placed within convenient location from where candidate can refer research data ! Ventilation and lighting arrangement in laboratory is made sufficiently fresh ! Emergency access is provided in laboratory for easy entry and exit !

All the instruments are labeled and their calibration record is available with lab ! Registers mentioning uses of these instrument is updated till previous day of work and anyone can cross check the same with respect to any random report prepared on any of the earlier day !

If there are some experimental arrangements being done and if some live study is going on , that area is specifically barricaded by placing red stickers and only authorized people are allowed to enter in that area !

The chemicals used in the lab are fresh and within the expiry period ! Their purchase and disposal record are easily available ! The frequency of calibration of different laboratory instrument is displayed on dash board of laboratory and there is no miss out in calibration !

The internal record of number of visitors visited your facility is available ! The photographs of visits and the memento shared during visit are stored in soft files !

Wherever necessary , one or two inspirational quotes are displayed for instant encouragement ! The laboratory area has good green cover and place for safe parking ! The internal roads are excellent and candidate can move there with the cycles or bikes and cars !

Security gate and labs internal area are connected with CCTV camera till entry point ! No unauthorized access is allowed during research work ! The tiles and flooring of lab is firm and at no place there are chances of slippage !

So , how will you feel when you enter inside such a world class research laboratory facility ? It will give you altogether different kind of experience !

18.3 Sophistication in handling research issues :

Friends ,

When you are doing any research , then there will be existence of some issues associated with that research topic !

Even though ,your facility is world class , still you will be refrained to use many facilities just because of certain laboratory protocols ! It doesn't mean you cannot use it , but it does mean some experienced scientist must accompany you when you are using that high end machine for your research ! This is what meant by sophistication in handling research issues !

Courtesy: Malupeca , Pixabay

Every research facility trains their student about handling high end machines ! When particular student achieves ease of operating that machine , he may be given charge of supporting another new student while using that machine ! This type of co-ordination makes fair use of that machine and when the other student become use to it , he is also given the independent opportunity to use that machine !

Second thing , if there are certain knowledge acquisition related issues during research like someone is copying someone's research with making small changes , such type of issues are also handled by your research guide with appropriate sophistication !

The reputation of research institute is of paramount importance and hence if such type of soft issues are noted , individuals are counselled to follow the laid down system protocol and in case such repeat experience is noted then it may hamper their research progress and they may be given some type of research break till the required discipline is restored in the system !

18.4 Sophistication in maintaining the transparency in research work :

As we have earlier seen, research work is highly important job for nation development and hence respective national agencies always remain in touch with prominent research institutes ! Suppose something undesired is happened inside the facility because of new approach to research and if high level enquiry is initiated, then how sophistically your research guide handle that issue is seen by everyone !

Courtesy : Pbear , Pixabay

First thing, when every small instruction is given in documented format, then it's just an exercise to back trace to relevant proof, put in front of enquiring authorities, share your additional explanation for their questions and sign the final closure report of that enquiry ! If the incidence is severe, either stringent protocols will be laid down to avoid such type of repeat experience ! If the incidence has deep impact on research philosophy, then certain hierarchical changes will be done to ensure right leaders are leading the research teams ! All such high-level decision making comes under sophistication ! The action will be taken and relevant person will be given detailed information about that action and decision !

18.5 Sophistication in regular communication related to research :

Research institutes are institutes of high repute and there is protocol for everything ! No one can shout on each other , no one can scold each other , no one can harm each other ! There is specific protocol for ethical code of conduct and everyone is supposed to follow it with care and caution !

So, your research guide and you are supposed to communicate gently and politely in such a way that it does not disturb to other researchers working nearby ! Such facilities are known for peaceful energy centers !

In next step , we will see ,research guide as presenter !

✱✱✱

MULTIPLE-CHOICE QUESTIONS

1) Which of the following is latest sophistication in internet technology in India ?

A) 3G
B) 4G
C) 5G
D) All of the above

2) Which of the following is latest sophistication in area surveillance ?

A) Close Circuit Camera Network
B) Drone cameras
C) Field Video graphy
D) Satellite Surveillance

STEP 19 : GUIDE AS A PRESENTER

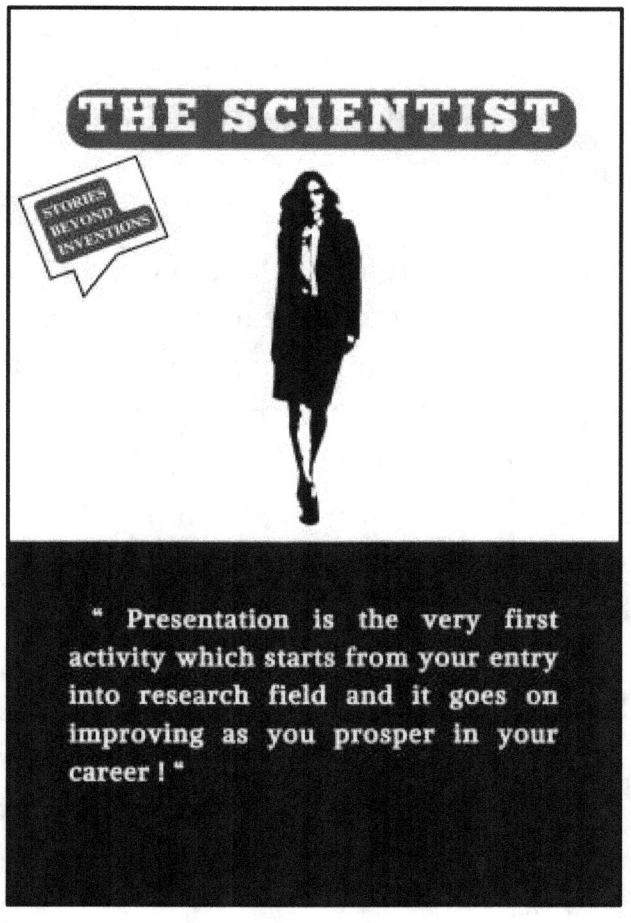

Image Courtesy: Open Clip Art , Pixabay.com

19.1 Introduction :

Friends ,

In this 19 th step , we are moving towards concluding side of scope of research guide in your research work ! In this discussion ,we are going to consider the role of your research guide as a presenter !

Presentation is considered as the end stage activity of every research work ,but this is absolutely wrong approach to look towards presentation ! Presentation is the very first activity which starts from your entry into research field and it goes on improving as you prosper in your career !

What is the difference observed in a cricketer playing for state level , national level and international level ? Is there any difference between the game played at these three levels ? Is the prize of winning same for all three stages of the sport ?

Friends , this simple example reflects the role of presentation in your research performance ! When you are competing with international research institutes which have phenomenal record of inventing new thing rapidly , you also have to understand the presentation skills developed by these students and the work done behind that level of perfection in their work !

19.2 Entry Level Presentation Skills :

Courtesy: GDJ, Pixabay

What is the difference you observe when you enter inside a temple or inside a playground ? When you enter inside a temple , the cleanliness and peaceful environment make your mind stable and happy ! The environment brings positive vibes to you and you feel totally content when you visit any temple or worship center ! This is the effect of positive energy created because of clean and clear atmosphere !

When you enter lush green playground with beautifully rolled out pitch with marking around the crease area , set up of stumps at two ends , umpire is standing in front of your batting stance and people are arrived with great enthusiasm and energy , how does you feel ,looking at this electrifying atmosphere ? Its gigantic feeling and this inspire you to score big and make your team ultimate winner against any opponent who is also playing to win only !

So , does the temple give its initial vibes on day 1 of its opening or later part also ? Is the cricket ground is made ready just for that match or it is ready at any time ?

Friends , this is what the requirement of presentation skills at entry level of research ! Before your arrival to research institute , your research guide has to ensure adequate research facilities are available in completely working condition ! Even if you start your work within less than a weeks' time , everything needs to be ready !

They say, first impression is the last impression ! When a highly qualified research grad enters your institute and when he observes how neat and tidy the environment and lab space is, they immediately like that environment and ultimately respect that place and carry out their work with strong research intention !

So, when student enters in such a sophisticated environment, naturally they behave according to group behavior which is studying over there ! This behavior attributes and culture of the institute is set up by team of your research guide and the leadership serving for that organization relentlessly !

19.3 Mid-Level Presentation Skills :

At this stage, every research guide, shares their student critical requirements of presentation during experimental work, writing work and public presentation work ! In this step

Courtesy: OCA, Pixabay

, the activities are arranged in such order that they can easily understood by the reviewers and hence further research will become easy to understand !

If you understand to notice this presentation requirement let us understand a third-party inspectors visit to your research lab ! At early stage , you have to first take their appointment by explaining the details of your visit call ! These details will be verified by the inspecting authorities at their record end ! If the prior information is registered with them as per protocol for start of project , they will give you suitable appointment ! If the project idea is not registered , they will ask to fulfill the balance requirement ! Once this activity is done , they will come to visit your lab on given date !

Before their arrival to your firm , you have to keep research work ready in all respect ! If there is experimental review , you have to keep that set up ready with all required material handy to access ! You have to keep research observations ready so that they can easily glance through it ! You have to keep additional set up ready , if first set up is rejected for some reason and you have to again show it with reserve set up ! If this is not done , they are not going to wait for 1-2 hour ! This is what the presentation skill required at middle level !

Then they will review your research file and may ask you to present on display screen ! For this you have to make presentation of your research work and it must finish within 2-3 minutes from your side and 5-7 minutes to be given for their interaction with you on research topics ! When all such type of preparation will be done from your side , then only the reviewer will okay it within fraction of second and thus your work will get required nod from reviewing authorities !

This is just one simple example of preparation for one visit ! In three to four years research tenure , you have to present your work to many prominent scientist and researchers to seek their views and suggestions ! Your research guide will also share your valuable work with them and in such sharing your work must be presentable and correct !

This presentation is important to look at your work as the work of a good researcher ! A good researcher has total clarity in his research topic and he systematically align the content , experimental findings, formulation and tables with graph presenting given scientific phenomena ! In today's digital age , one can present their research wide a well described video prepared by you and professional artist adding creative value to your research work !

19.4 Presentation during finishing touches :

In three to four years' time , your research guide prepare yourself for this final moment ! Without the investment of joint effort from day 1 to final moments of presentation , this presentation skill become difficult to build !

So, a research aspirant wants a path breaking invention while the research guide wants their tradition of exemplary researchers must flourish to next level of human intelligence ! Its dream of every research guide that his future research batch must be sharper than their previous batch ! His experience of coaching is improving constantly and he wish his guidance must make things more easy , more sophisticated and more live issues will be researched in his lab !

Courtesy: CFV , Pixabay

It's the contribution of research guide that their organizations are known for carrying out cutting age research to shape up the world in more simplified way ! When your laboratory is carrying out most advance level of research , many visitors keep visiting and adding value to the research observed ! Its mutually respectable forward looking cultural exchange and at such level , presentation skills need to be at par !

When global researchers come to invent something beautiful for humanity , the presentation skills need to be excellent ! At this stage you are not representing yourself but you are representing first your highly renown research institute and second , the country which is making such joint work possible ! Because of progressive government's active support and encouragement , such type of joint exercises rarely happens ! The joint efforts and involvement of government indicate there is global unison approach to look at the live problems and when such issues will be resolved ,the global population will benefit from it and this way many international challenges will be tackled comparatively easily ! So, a fluent English , apt use of presentation tools and a warm attitude to respond to queries is what desired in international presentation opportunities !

19.5 Presentation at professional level :

When you observe , someone working at high profile position after 30 -40 years of dedicated study , you notice the person is very very sharp in his thinking , his decision-making capability is excellent , his approach towards looking around problems is unique and his interpersonal skills with people is just very very simple yet very much supportive !

How this transformation of a school going kid to a professional master happens at the age of 30-40 years ? This is the end product of well-educated atmosphere creation and maintenance ! In the initial phase , you need extremely powerful efforts to set the things in right order , but when your set up and efforts starts functioning , highly talented researchers keep qualifying through such research institutes ! When these students join prominent organizations , they add their valuable knowledge to that organization ! This stand enhances business atmosphere and more people working with such leaders become capable to resolve various live issues !

The presentation skills developed at school and college level boost the development of presentation skills required at research level ! Hence in the very very initial stage of schooling , participation of student in various group discussions , stage performance is necessary to feel confident and determined at any level of professional opportunities ! Your research guide just polishes your presentation skills build over these years of dedicated study and personal development !⊛⊛⊛

MULTIPLE-CHOICE QUESTIONS

1) Which presentation skills are must to prepare a research project report ?

   A) Detailed theoretical coverage
   B) Attention to all details
   C) Mention of Practical applications of research to society needs
   D) All of the above

2) Which type of presentation is much desired for a research institute ?

   A) Modern Facilities connecting to every research element conveniently .
   B) Peaceful ,neat and spacious laboratories .
   C) Quick and intelligent research report review fascility
   D) All of the above

STEP 20 : GUIDE AS A GENUINE CITIZEN

" However tough the research work may be , research guide needs to provide the confidence to their student that there is always light at the end of the tunnel ! Research guide always need to tell their students ,when going get tough , tough gets going ! Research guide must talk with their team about lighter happenings in life !

Image Courtesy: U_FG , Pixabay.com

20.1 Introduction :

Friends ,

Welcome to this concluding discussion of typical attributes of research guide ! Till now ,we have discussed ,what it takes to be a researcher which is followed by the attributes of research guide ! The next 10 steps will discuss the research carried in some of the well-known research fields !

As we have seen earlier , research and development is a nation building activity and it has keen role of government ! Without government's involvement and policy support , your research will be just an unauthorized invention ! This is because , the scope of any research directly affects public life . Once public observe ,some new thing is found out and its extremely attractive and available in lower price , as per natural human tendency , public try to own that thing and this make consumption of that invention a public interest issue !

When such unauthorized use causes some public injury then respective person make a formal complaint with police station which are representative of Government from governments side to protect the people in that respective region !

So , when police take note of that case , they file a FIR (Field Investigation Report) which consist of typical raid at place from where the invention is sold to people , the agency or person responsible for that invention and legal remedy to investigate how this invention is reached to people without knowledge of police and their information network !

When such shocking revelations are done , the matter reach to courts where Honorary judges refer to the case laws and listen to both sides ! After completion of due process of law, the culprit is punished and to avoid such incidences in future , required law amendments are done to cover the new version of cheating and looting people ! This is how an unauthorized invention proves as harmful for society and hence every research work needs formal government permission before it is published in any scientific societies ! Government of all nations are supreme care taker of their people and hence no scientific body has higher authority than Government however intelligent and smart they may be ! Because Government means faith of people ! The matters which are related to health and money are highly important matters and hence government's intervention is must on these topics !

Courtesy: Bored, Pixabay

On the other hand , when there is formation of national science board under which all type of research work is registered and approved after adherence to due protocol of research , such work receives governments approval from competent authorities and thus any public can use those products without any worry !

Second thing , in case some issues are arises in product made from that research, government will have well planned inspection plan to such facilities where that particular producer has to show products made from that research work ! So, there will be system certification to ensure the products meet general guidelines of safety , quality and environmental protection ! When such adherence is observed , Government approves those products before selling to public either by themselves or with the help of technical agency representing government who is expert in such type of product creation of private and public use !

So , if out of one year's production , if 11 months production refer safe output and 1 month's output shows customer complaint , then the all production in that period is verified and if there are major quality deviations , same are noted and the root cause of customer complaint is corrected in the system with applicable penalty to respective organization for breaking working rules laid down by Government as per governing law or regulation for that product !

Friends , this is all about ,the introduction of research guide as a genuine citizen ! He is certifying people who are going to be next nation builders and their contribution will have major impact on todays and tomorrow's society ! So , let's see , which personal attributes are essential in your research guide as a genuine citizen !

20.2 Freedom of Expression :

Learning is a free-flowing experience which consist of knowledge and ignorance ! When students are doing self-study , they have to put more efforts to simplify the understanding of key concepts and their interrelation ! However, when a well-qualified and experienced research guide is present near you , you can ask any type of doubt to clarify your concept knowledge ! It is required that at research level , you must possess basic fundamental knowledge but it is not that much essential when your research guide is absolutely free minded ! He will explain you that concept again and again , so that your

Courtesy: OCA, Pixabay

basic foundation will again become strong and stronger !

Always remember , the sum which seems to be difficult in standard 5th is become a few seconds job in standard 10th ! The sum which look tough in standard 10th become a jolly work in standard 12th ! The sum which looks difficult in standard 12th become a cakewalk in fourth year of engineering and so on !

So , when you ask some doubts at research level , it's absolutely fine , your research guide knows about it partially or fully and he will guide you about how to approach this unknown ! This is because he has gone through such mutual learning experience over the number of years of this professional service and they have always increased the knowledge base of their subject expertise . Because of this knowledge surplus , they can resolve your doubts easily !

20.3 Equality of research opportunities :

Research is highly competitive and lucrative career option and hence there is tough competition to excel and grow ! In such field , your research guide needs to provide equal opportunity to every researcher !

Courtesv: CFV. Pixabav

Suppose there is team of 20 working researchers and if there are 5 research guides , then every research guide must coach to minimum 4 students to make required knowledge experiments suitably reviewed when they are being done !

Second thing , there has to be adequate research equipment's so that there should not be a second waiting for its usage ! The time invested in research work is of high importance and many researchers choose research subjects which holds either national importance or which are related to national defense ! Hence , every research guide must try to provide all required resources to every researcher without any wait !

20.4 Co-operation and conducive atmosphere creator :

A highly motivated team of scientists when works in unison , the world class inventions are researched ! In number of space missions , collective efforts of scientists and their guides has shown nothing is impossible if you are following fundamental rules of science ! Every research guide is responsible for creation of such conducive and mutually co-operative atmosphere ! In the end , little bit time slag is observed ,but every researcher put his total efforts to complete the research in given time ! When fundamental research logic is achieved , the next step of inventions are experimental steps where success is bound to happen after multiple relentless efforts ! A research guide must have knack of this research breakthrough !

20.5 Human Values :

However tough the research work may be , research guide needs to provide the confidence to their student that there is always light at the end of the tunnel ! Research guide always need to tell their students ,when going get tough , tough gets going ! Research guide must talk with their team about lighter happenings in life !

They must discuss normal life events and find some time to hang out outside for a trek or some mind-blowing refreshment visits ! Going to hotels for multicuisine food experience and networking in breakthrough parties also need to be done !

Work -Life balance is need of every person and researchers are also no different community ! They are most sophisticated intellectual talent available in nation and hence they need to take care of their brain , mind and soulful heart as well as their dear ones ! Invention made by scientist is always remembered but what may happen during experimentation phase of invention is rarely come to surface !

There are experiences of some type of causalities during typical high-risk research work ! Inventions related to radioactive and nuclear uses material possess ultra-high risk of radiation leakage ! Experiments related to study of fission energy of chemical also has risk of untimely and uncontrolled explosion !

Experiments related to food stuffs also has risk of inhalation of unhealthy composition ! Drug and medicine preparation also has health risks because of continuous exposure to toxic chemicals ! So, this is the general work life of every researcher ! They have to face challenges in very very early stage before people start using that invention !

That's why a research guide must be a person with vast knowledge but exact wisdom to share that knowledge with the best candidate contributing to the field of research ! He must be a person with high orientation of scientific thinking and its application to direct and indirect human life ! These researchers become major value adder to any field in which they are serving ! After all its social and nation building work and hence commitment to final duty of public service is utmost important !

In next step , we are spreading light on path breaking research ! ✸✸✸

MULTIPLE-CHOICE QUESTIONS

1) **What are the consequences of unauthorized research ?**

   A) Default Public safety concerns
   B) Direct effect on people's life style
   C) Gross violation of financial disciplines
   D) All of the above .

2) **What are the benefits of Government aided research work ?**

   A) Research is considered as important service for nation building .
   B) Research creates international identity for country .
   C) As research is government aided , its benefits are transferred to public directly
   D) All of the above

STEP 21 : PATH BREAKING RESEARCH

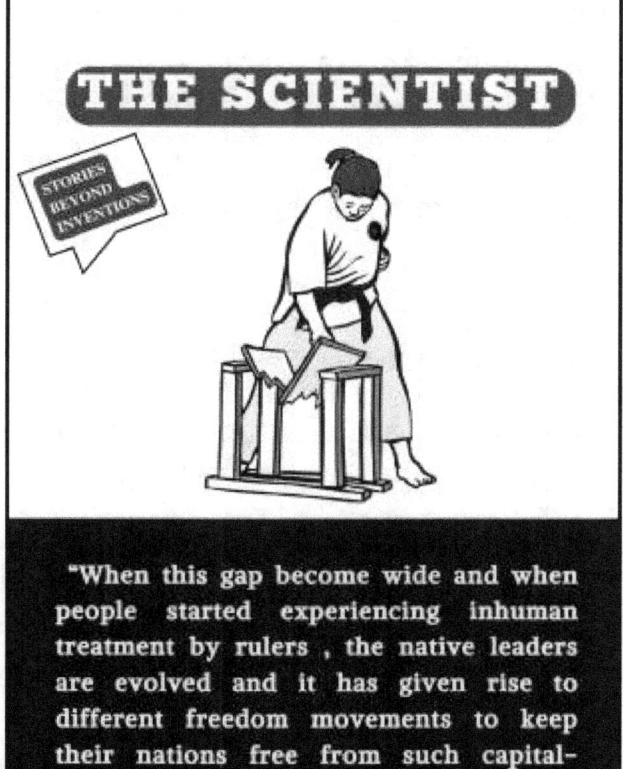

"When this gap become wide and when people started experiencing inhuman treatment by rulers , the native leaders are evolved and it has given rise to different freedom movements to keep their nations free from such capital-oriented rules where their national treasures were looted like nothing ! "

Image Courtesy: Clker Free Vector , Pixabay.com

21.1 Introduction :

Friends ,

Welcome to the last section of this book 'The Scientist – Stories beyond inventions ' ! This section will spread light on the impact of path breaking inventions on the society ! How society was struggling before a specific invention and how many changes brought up by inventions after their commercial release !

21.2 Path Breaking Research :

Basically , path breaking research signifies the importance of research for humanity ! There is typical path of living normal routine life ! When scientists all across the world started their research work to understand the nature around and its magical phenomenon , the scientific evolution of human culture started developing !

So , what was before invention of wheel ? People were doing their work by walking or running ! They were using few animals for their transportation needs ! When the invention of wheel is occurred , then the path of transportation is changed phenomenally and now we are living in an area of bullet train ! So ,just see how much radical path braking research is happened in understanding the core fundamental concept of wheel !

Courtesy : OCA, Pixabay

On the other hand , invention of fire has ignited deep sense of research amongst the scientist and they found out various material which has high calorific value ! When they burnt such high caloriphic value material in presence of air to melt metal ore like iron ore , they find out the method of iron making ! When further research on iron making was done , the technique of steel making is evolved ! On further level of research , the method of alloying was invented and trailed and this given rise to making of stainless steel and other high alloy steel ! Once the melting point and freezing point of material is determined scientifically and recorded in the properties register , scientist started using that material for high temperature and low temperature applications ! This is how basic element's periodic table was invented and material is classified into acidic and basic class of their chemical affinity ! So, in this tenure , one invention was literally supporting other invention and this research work jointly added to string of path breaking inventions in many fields !

When research was carried out on properties of light , the inventions like lens and mirror are occurred ! The manufacturing of glass was also an important invention ! When the eye wears are invented , it given comfort to person suffering with vision issues ! The spectacles proved as supported vision to elderly people and nowadays almost after attaining age of 40 years , people are using eyewear's for far or near vision acuity issues !

The lens technology was further utilized to develop the camera for clicking image of an object ! This was again major research in the history of mankind ! The photos taken with the help of camera are developed in lab ! With the help of camera roll , images were clicked as negative and when they are developed and their impressions are taken on photo quality paper , the photographs were produced ! Later , the camera technology is further researched to invent video shooting camera ! This camera was catching live moment of any incidence and hence storing memorable memories for long time become possible because of invention of video camera ! How many brands emerged with this invention is a well-known fact !

The video making technology has given birth to today's cine industry ! Earlier plays were presented live in theaters and public locations ! With the technology of movie screening ,people started enjoying the movies in theaters and that is also quite frequently !

You shoot a film once and show it to million people by making copies of that original print ! The invention of copiers and printers thus supported the easy distribution of movie prints !

When research is focused on printing press and people come to know that the paper can be made from typical species of trees , they started making paper from these tree wood ! By that time ,printing press was also developed where letters and numbers and pictures were printed on papers fed to the machine ! This technology given rise to the business of publication ! Publishers started printing newspapers, books , literature , wedding card throughout the world for basic knowledge transfer and interpersonal communication needs of the people ! The business of publishing is still running although some lateral competition in the form of media and digital media is occurred , but still people read print material because biologically it is less stressful to eyes and it also allows you to read as per your convenience and without using any electricity !

So , this is the small introduction of path breaking research carried out by veteran scientist and how they changed the old world to current level of sophistication ! In this section , we are going to discuss these research breakthroughs category wise ! Now we will see , what are the impacts of these path breaking research for society !

21.3 Impact of Path breaking research :

The first and foremost impact of path breaking research happened on the rise of industrial revolution and huge scope for employment and living a decent life ! When inventions are commercialized and sold across the continent the countries which were not able to manufacture it , started consuming it by paying its price !

This given rise to international trade and hence countries who were pioneer in the field of scientific research started manufacturing products and started distributing it across the continents !

The industrial growth of one nation thus fostered the process of modernization and living an ultramodern lifestyle to seek more comfort and more financial freedom !

On one side of the world , there were people who were not literate enough to learn the scientific manufacturing process and prepare products of high quality ! On other side , there were people who were excellent in creating these products and able to sell it across the world ! This made the typical financial divide and countries with developed economy and under development economy are created ! There was another world which will living in tribal communities and scientific development was miles away task ! The colonization period of history typically reiterates the fact that wherever colonization occurred, the locals were traded these goods at a selling price and ample raw material available in their territory was made available for manufacturing in their home state !

This type of dealing kept the intellectual part of creation with the scientific community and receiver community just received final products at a sales price ! On this basis, huge financial empire is set by colonization process and later the same people taken the charge of provinces to run the princely states with their own rules of governance !

Hundred and two hundred years of slavery was started and later history of race and ethnicity is well known to everyone ! So, with the settlement of capitalism, inventors become rich and consumers become poor !

Courtesy : Star Glade Vintage, Pixabay

When this gap become wide and when people started experiencing inhuman treatment by rulers , the native leaders are evolved and it has given rise to different freedom movements to keep their nations free from such capital-oriented rules where their national treasures were looted like nothing ! This phase of modern human history is also considered as one of the rapid stages of industrial expansion !

21.4 Globalization :

When countries become free from colonial rules , they started developing on their own with the help of infrastructure facilities built by earlier rulers ! The previous traditional infrastructure plus infrastructure built during colonial rule is taken as input for national industrial development and elected national government started establishing education institutes and commercial facilities !

The industrialists with high degree of nationalism , started their industries and manufactured parts like steel , railway, agricultural products , health products and stuffs which are useful in daily life ! This phase of national industrial development was comparatively slower and people has to wait for years to receive the parts manufactured in national facilities !

In the 1990's , the concept of globalization and liberal economic steps were taken in India and with this Indian market become open to international establishments ! Now ,by following due course of law , companies are allowed to set their plant in India and appoint the local talent for their regular operations . This boosted the rate of employment and also changed quality of life of people employed in such organizations !

Courtesy: OCA, Pixabay

Indian products are also traded in other nation and from these places foreign currency started coming to India ! India is basically agricultural prone economy ! After meeting domestic consumption limit , high quality Agri products are traded and foreign currency is received !

The typical agricultural reforms ensured the production capacity is improved and farmers are getting quality input to do efficient farming ! More such liberalization impacts are noted and India is standing in the list of some of the important global economies experiencing promising growth !

So , again the research is proving itself as mean of nation building and continuous improvement to take care of your people and utilize your resources carefully ! The country which focuses their technologies to use resources optimally naturally saves valuable resources and over the period of time their cost goes on increasing which again adds to foreign exchange when these resources are traded to other country ! Hence one must try to develop new techniques with which they can save natural resources available with them !

In the next chapter , we will see the examples of impactful research ! ✷✷✷

MULTIPLE-CHOICE QUESTIONS

1) Which of the following invention can be called as a path breaking invention ?

   A) Computer memory hard disc
   B) Internet Technology
   C) Wi-fi network
   D) All of the above

2) How many fields get impacted with one path breaking invention ?

   A) Technical Field and allied businesses
   B) Social Fields and people's behavior
   C) Political set up and decision making
   D) All of the above

STEP 22 : IMPACTFUL RESEARCH

Image Courtesy: Sherline , Pixabay.com

22.1 Introduction :

Friends ,

In last discussion , we have seen ,how path breaking research is done and how humanity is benefited from that research ! On the other hand , there were few disadvantages observed when given instruction of invention uses are not adhered by public ! No research is perfect at very first stage of its discovery , when iterative versions are created with small small development steps ,then only it achieves its final form with most desired perfection !

Since , being the manual work , there is some scope for imperfection and that's why every scientific invention has to be used within limits specified by its inventor !

In this discussion , we are going to see how impactful research make visible changes in society and how other business network get created because of one scientific invention ! We are also going to see ;how negative impacts of invention are observed in society and what type of controlling mechanism is necessary to take care of such effects !

22.2 Impact on lifestyle :

Any research when become an invention and when it is produced in large scale , it makes impact on people's daily lifestyle !

Courtesy: Jozefm,Pixabay

What is the impact created by smart phones in your daily life ? Twenty years ago , smart phones were rarely available in India and very few people were using it for their official and personal work ! The prices of using those services were very very high as well as the mobile network which is required to transfer the telecommunication signal was not present !

But as soon as this infrastructure development is done by mobile service providing organization , in a flash , people opened mobile shops , mobile repair centers and prepaid-postpaid subscription outlet ! They employed staff over there and started serving for public !

In the initial few years, mobile bill was sent to designated address with all call details and tariff charges, people used to visit nearest outlet for payment of mobile services, if there were few billing issues, same were corrected at such centers ! With further level of research in this field, high speed internet connection has made revolutionary impact in people's lifestyle ! Starting from 1G, moving to 2G & 3G, mobile internet was used for normal surfing purpose for fetching required information ! But when 4G and mobile applications are evolved, the world got a virtual space where financial transactions can be easily done at the click of a button ! The world cannot be made so simple, but yes, telecom research and development made this task feasible !

With the inclusion of service providing applications a bridge is connected between creator of service and consumer of service ! Once the typical service is booked on the application, both creator and consumer were able to look the progress of that service ! This transparency made such service very very impactful ! People used these services and paid its service charge either instantly or after getting delivery of the service !

The rise of e-commerce using mobile technology is the next step of upgradation of lifestyle of people ! Because of pre booked door step services at reasonable price and offers , many people started purchasing same items form online market ! The benefit of online transactions is seen during pandemic times when all physical activities were at halt ! Now if someone think , what would have happened if digital world was not available at that time , the person will not able to imaging the chaos and frustrations which might be created because of uncontrollable sense of sudden urgency !

This is how technology and research work provide you a simple and better option to deal with tough times and enjoy your life as it comes ! Positively and innovatively !

22.3 Impact on Financial Movement :

When basic economics of demand and supply are considered , every useful invention creates huge demand pool and hence high scope of employment generation ! Skilled people have to produce those products by using cutting edge technologies and they have to supply it to your domestic and export market ! With this supply, money comes to your nation and thus your nation become wealthy and rich !

What your nation can not achieve if they sincerely decide to develop it with accumulated wealth ? Apart from the degree of undesired and unhealthy corruption in the society, still your nation can do so many things for your people !

Courtesy : OCA, Pixabay

They can construct roads and highways for people for easy and safe transportation, they can support professional courses and can respect the contribution of teachers at all levels, they can improve health services network for people so that critical illnesses can be cured within reasonable fees, they can improve their foreign relations and may invite big brands to set their branches in your country so that more employment can be generated , this is rise boost economic activity in your country and your foreign exchange rate improved because of rise in per capita income and your gross domestic product !

With high wealth , you can maintain social equilibrium in the society by supporting producers in taking record farm yield which can lower the per kg cost of their product !

When there is adequate food supply is available in your stores , the prices remain under control and inflation is also avoided beyond intolerable levels ! Low levels of inflation make people buy more and thus it improves sales of those products ! With more sales , more profit is attained and thus both consumer and producer maintain a healthy professional relation with each other !

When there is inflation , people don't buy more and they tend to reduce consumption or they may look for cheaper option ! This increases cut throat competition between producers and sadly the quality of product keeps deteriorating for business survival !

This is negative impact of demand -supply equation and hence people should never compromise on quality of products ! When you accept substandard quality for less price , the manufacturer keeps producing further lower grade as people keep buying the same ! Instead, if people buy good quality product at higher price , the product retains its market importance and hence people consume what is best for them at all time !

If you see , how a little saving of 10000 or 20000 affect you in the end , one can understand the conditions of frequent repair and work stoppages , injuries to person using substandard machines , extra time required because of lack of fast working modes and features because you need high strength material when you are doing something very very fast ! If low grade material is used , it will fail after two-three attempts of fast working ! You cannot complete your orders in time ! This gives rise to professional mis-understanding and hence you need to take care about using product having good quality for your daily business activities ! Although it is costly ,but is saves your so many other costs !

22.4 Impact on Health :

Health impacts in every invention are both positive and negative ! We will first discuss the positive impacts !

Earlier radiographic imaging technique was not available ! Hence detection of internal body defects was not feasible ! Doctors used to apply their knowledge and experience over the years to cure the patient from such internal diseases !

The technology of radiographic imaging improved the quality of diagnosis and hence detection of root cause become simple ! Now doctors are able to see the size of problem and its subsequent growth ,so that they can avoid it by specifying required medication programme !

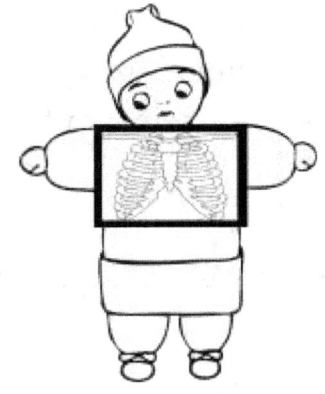

Courtesy: CFV , Pixabay

Same is the thing for ultrasound technology ! This technology is used in child birth and growth assessment ! With this technology carrying out safe baby birth become possible and monitoring of both child and mother become easy and accurate !

The inventions in micro biology studied different type of organisms and their lifecycle to cure diseases which are caused by spread of such viruses ! Because of understanding their pattern and nature , formulation of protective drugs and medicine become possible ! Scientists worked with smart approach to vaccinate child in their early age so that such type of infections should not happen in later stage of life ! This was the most impactful research to take care of future generations in advance !

22.5 Impact on those not using any invention :

This is the other side of invention application ! When the scientific development was started in western part of the world , the eastern part was not aware of such development for quite a long time ! People were living as per traditional customs and rituals which are set in the society by divine saints , philosophers and social activists !

There were many misconceptions about using inventions as using the natural prowess and power for personal use was considered as not healthy sign for future generations ! This approach was fairly relevant till the population of globe was limited ! When world noticed rise in population , with available technology ,they started mass production at huge scale so that basic needs can be met at fair pricing of commodities , products and services !

But still there are some tribal and ethnic communities which are not using any modern invention ! They are living a simple and healthy life in presence of nature and they are making their earning by selling either agricultural products or some natural fruits available in their jungles !

So , if you compare a modern person to tribal person , both are fulfilling the basic needs of life from the same nature ! Only difference is , the modern man is in pursuit of different levels of consumption while the tribal person is happy with what he is ultimately receiving from nature ! So , the impact of research is only felt if you use modern products and services ! ✳✳✳

MULTIPLE-CHOICE QUESTIONS

1) Which of the following is the most impactful invention for global financial system ?

A) International credit-debit master card
B) Mobile Payment Applications
C) Bit coins and virtual currency
D) Multinational Banking

2) Which of the following invention is most impactful for health sciences ?

A) Influenza pills
B) Cough Syrup
C) Insulin injection
D) COVID-19 vaccine

STEP 23 : RESEARCH FOR DEFENSE

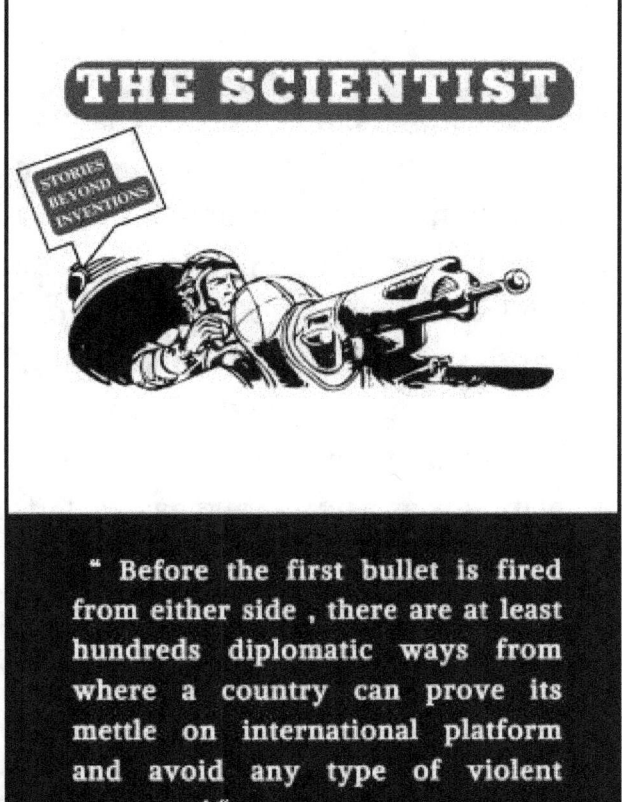

" Before the first bullet is fired from either side , there are at least hundreds diplomatic ways from where a country can prove its mettle on international platform and avoid any type of violent response ! "

Image Courtesy: Open Clip Art , Pixabay.com

23.1 Introduction :

Friends ,

We are now slowly moving towards different fields where research contributes to exceptional proportion ! One of the such field is defense ! Defense means both internal and external defense ! Earlier meaning of defense was only related to human attacks occurring from other parts of the national boundaries but in 21st century, definition of defense is broadened and now defense means protection of not only your people and property but protection of your constitutional value system that provide freedom , equality and brotherhood to your descent citizens as a national treasure ! Efforts of Protecting constitution from internal and external attacks or attempts of attacks ! So ,let's move to different fields where research for defense is directly applied .

23.2 Traditional defense :

Every country has their own national defense system ! The traditional defense mechanism work in different methods for army , navy and air force ! So , you have to constantly orient your efforts to strengthen these sectors of your national defense !

Courtesy: OCA , Pixabay

Research done for development of army includes inventions of tools and tackles that provide next generation arms to the brave soldiers ! Every soldier is priceless and so as his contribution to national service ! So , the organization which are strengthening army has a single goal to train your soldiers to cutting edge target execution !

There can be improvement is field defense strategies as well as improvements in working condition at challenging locations ! If your soldiers are serving in safe place , chances of that area's safety increase ! If the soldiers are serving in tough condition , you have to provide them all tools and tackles which will take care of that area as well as that soldier !

So , what a general defense material includes ? The newly designed war explosives , swords , guns , tanks , missiles and such weapons which has mass destruction capacity of high accuracy to reach at a place before your enemy could ever think ! Wars are won by strategic thinking done near war footing ! If that thinking is exceptional , less soldiers need to get injured and martyred for nation !

Before the first bullet is fired from either side , there are at least hundreds diplomatic ways from where a country can prove its mettle on international platform and avoid any type of violent response !

The next section is research for navy ! Navy protects your waters and send you alarms when they see suspicious movement around international sea lines ! The major chunk of defense in Navy is done with the help of advanced navigation systems and hence they need to be upgraded with advanced navigation systems ! The giant war ships have different role in Navy's defense ! Especially for providing landing place for Airforce fighter planes !

Apart from that , navy carry out many rescue operations when you have to bring your people back from troublesome places around the world ! Navy also saves people who has regular work in deep sea zone and who doesn't have any saving mechanism of their own ! The current development in national navy's shipyard is testimony of the fact that Navy is getting what is required to strengthen the overall capacity and capability of navy !

Third defense research is done for air force ! Air force research consist of development of high-end fighter planes ! Development of high-end missiles with accurate target sensing capacity ! Training to pilots to carry out operations in severe working conditions !

The field of aerial communication is also improving and with inclusion of computerized signal system and also sharing of different sky lines , one can easily trace if enemy or any other air craft is entering your airline limits for any unprofessional purpose ! The necessary response is immediately given by your air force defense !

Let it be use of helicopter, use of fighter planes or passenger planes, air force has its own way of working and during war time, their contribution in attacking target place is immense! Because of this sharpness, target get hit in minimal time and operations are completed with maximum accuracy! The advanced target sensing technology provides accurate positioning from distance height!

23.3 Financial Defense:

Finance is lifeline for any nation and when their economy is attacked, the nation naturally lacks progress opportunities! So, what type of attacks comes into finance attacks?

Some nations are prohibited in doing international trade transaction on either side! This creates resource shortages internally and their final product also not get any external

Courtesy: OCA, Pixabay

market from where they can earn foreign exchange !

Second attack consist of constantly devaluating currency of target nation ! If you have 5 resources , in the international market ,based on your internal trade relations and mutual need , you decide the exchange rate of those product and services ! So , if nation is only procuring things from you and you are not buying anything from that nation , then the procuring nation may get that item at high price ! When there is mutual trade , the dependency factor made the price range within mutually agreeable level ! So, out of 5 resources , 2 resources will be sold at higher price , 2 will be sold at lower price and 1 may be sold at equal price ! So , this way ,nations can be targeted on international platform ! To avoid such economic attacks on nations , there is formation of regional co-operation and well-wishers' association just like SAARC and ASEAN !

Third economic attack involves charging excessive tax and custom duties for material coming from or going to particular country ! If the country doesn't have sound interpersonal relation with you and still want to trade with you , you do such type of trading at higher costs to keep your sales dominance ! There are many small countries , who purchase products from giant nations at comparatively higher prices !

This is the reason why developed nation become more developed and wealthier ! Price disparity factor is always in place in such type of international trades !

Stoppage of critical production of items for some time till high demand get created is another economic tactic played by few nations to prove their economic dominance over other nations !

So, to remain free from such type of attacks , you have to develop your internal economic framework conducive to internal growth ! You have to export products to countries that has descent economic and cultural relation with you ! You have to support each other in times of distress ! More the allies on your side , more the financial gains for your nation ! More the enemy against your nation , you have to feel several financial attacks from all such nations !

So, the people running the nation now need to take care of international goodwill creation and newer technologies to be used to protect your financial systems form internal and external frauds and suspicious transactions !

23.4 Health Defense :

In a highly interconnected world , the national health defense protocols and research done in this field is become an important aspect of research for defense ! Recent pandemic has proved this aspect quite clearly !

Courtesy : OCA , Pixabay

People has seen a deadly virus has entered human body system and it is transferred from nation to nation ! The spread was fast and to prevent this spread ,required immediate medicines or vaccines were not researched ! The virus was changing its nature and many mutations were seen during its presence !

So, scientist researched a vaccine which can protect from harmful effects of virus infection ! But by that time, lots of harm is caused to national employment and economy !

In earlier times also, such type of contagious diseases is occurred and with development of vaccines and preventive medicines such attacks are nullified !

So, what will happen in future ? Don't know to anyone, but you have to keep your research lab upgraded enough in line with international developments, so when challenging situation occurs, you will be able to deal with it accurately !

Apart from such infections, rise of particular disease in kids, youths and elders is also a challenging situation for health warriors and hence they have to apply their research logic to such situations ! When a breakthrough drug is invented, lakhs of people get their life saved ! Thus, nation achieves heath defense from such drugs !

So, it's quite clear that ' Health is wealth ' ! Protect your health and defend your wealth !

23.5 Cultural Defense :

As we have seen , attacks on constitutional values need to tackle with research done in various social sciences ranging from research in psychology , terrorism , narcotics , human trafficking , illegal transactions of human and animal organs !

These types of attacks also need sound defense mechanism ! One must know , how the illegal smuggling of sandalwood is nullified by police operations ! One must know how the precious metals were illegally traded inside nation and how police tracked it through their network ! Recently human trafficking and illegal organ transplant is becoming new way of internal risks ! Research done in such live social issues is proving as a simplifier and defense systems are making best use of available forensic expertise to crack the motive and impact behind such harmful deeds !

So ,as we have discussed earlier , research is a nation building work and if you have to remain safe and protected , you have to always prioritize your research activities !

MULTIPLE-CHOICE QUESTIONS

1) In which of the following sections research for defense for army must happen frequently ?

A) Arms & Ammunition
B) Soldiers Training and international war practicals with friendly nations
C) Various signaling systems and coding
D) All of the above

2) Which of the following area is looked after by research done for navy ?

A) Facilities is advanced war ships
B) Research in planning dangerous rescue operations
C) Research in co-ordination between Airforce -army and navy on the base of major warships
D) All of the above

STEP 24 : RESEARCH FOR BUSINESS

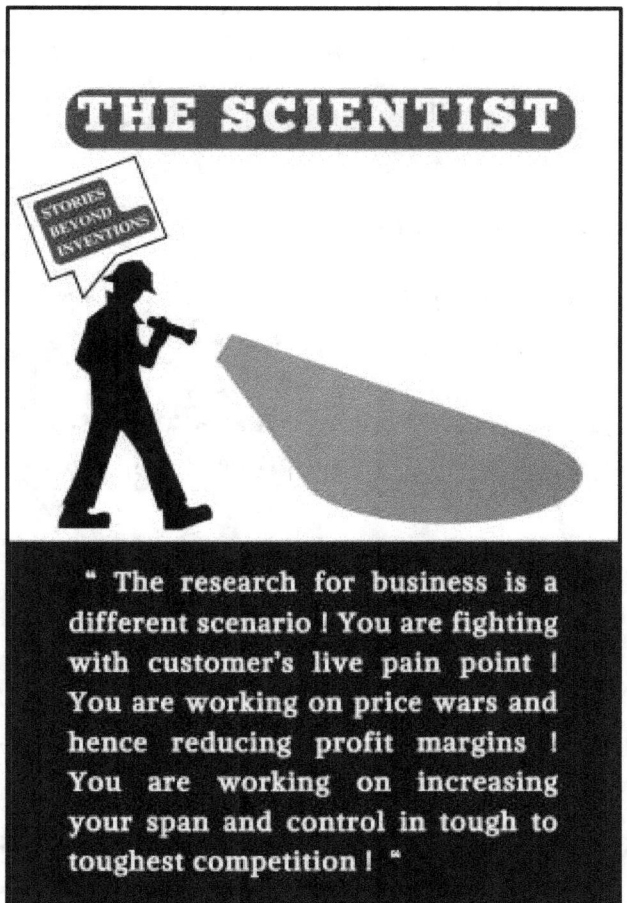

Image Courtesy: Clker Free Vector , Pixabay.com

24.1 Introduction :

Friends ,

We have discussed about path breaking research , impactful research , research for defense and now we are moving towards research for business !

How many time your country can be targeted by your enemy ?

How many times you need to launch critical missiles towards enemy's base ?

How many times you have to send your soldiers on line of control with large size troops ?

Friends , beside of regular patrolling and pre-determined field protocol , the need of use of weapons and force occurs only when there is attempt of attack from your enemy side ! Otherwise , you never attack anyone intentionally ! So , research for defense is ongoing work and when this work reaches its typical milestones , the field trails are taken and the strength of that particular section is enhanced but the weapons are never used until the formal order is given by government ! So, every country as per their defense vision , keep inventing new weapons and missile to stay in line with progress of other mighty nations !

The research for business is a different scenario ! You are fighting with customer's live pain point ! You are working on price wars and hence reducing profit margins ! You are working on increasing your span and control in tough to toughest competition ! You are fighting against high cycle time and finding avenues to reduce this cycle time to improve operational efficiency ! You are fighting with new government tax rules and how to set your prices tightly so that your business turnover attains its peak performance throughout the year ! You are fighting for developing suppliers at various workplaces and getting their best possible contribution to your business ! You are fighting for achieving a breakthrough invention which will boost your sales exponentially and increase your all options of becoming number one player of the business world ! This is where , research for business really matters ! In this discussion , we are going to see ,how the research for business contributes for overall business development and growth !

24.2 Business research to tap unknown market potential :

What is the most important task and duty of any business leader ? If someone ask this question to a new MBA pass out, his answer can be related to product development and process development ! Few brilliant students may answer market research and development of new product that fulfills market demands ! But if you ask this question to any seasoned business researcher, he will answer that he will try to create a market for his breakthrough invention !

Courtesy: Tumisu, Pixabay

Yes, creating a brand-new market for your product is the first and foremost duty of any business researcher ! When you create brand new market for your product, you are the first entrant into that market and hence you have all chances to gather the huge customers form that market until your competitor also enters into that market ! All the business competition revolves around this race of entering new market !

Who started telecom mobile services first time in India and what is their current market penetration ? Who started cable network services in India very first time and what is their current market penetration ? Who started the CNC machine manufacturing in India and what is their current market penetration ? Who started the electric two-wheeler in India and what is their current penetration ?

Now , let us consider ,the other side of business research ! Who are the latest entrants in the telecom mobile service provider and what is their market penetration ? Who is the latest player in cable TV market and what is their current market penetration ? Who is the latest player in CNC machining and what is their market penetration ? Who is the latest player in electric two-wheeler and what is their market penetration ?

Friends , if you are entering into product market as promoter ,then you get the 100 % benefit of market potential ! Other player may enter with you or may enter after certain observation period ! So , in the early period , you have to launch your new product in such a way that it has excellent quality , optimum price , quickest after sales network and easy access to designated offices for business deals and transactions ! In such a case , you get the most of that market and wins that market almost single handedly ! If you have seen the growth of telecom market since 2001 , there were only two mobile service providers and they were charging for incoming calls also ! In 2024 , it sounds funny , but it was the potential of that market at that time ! Slowly , the telecom networks are spread from metros to cities to rural area and now there is entry of additional 4-5 big players ! So , ultimately , the early bird advantage of this market is over and now organizations are doing business based on their traditional experience of retaining earned customer over the several years of customer service !

Same thing is observed for EdTech startups before pandemic and lockdown ! Providing customized online educational training courses was a new market and it was either subscription based or one-time payment based ! When these types of start-ups were started , it received slow response ! Basically, few students and professionals from urban area used these services ! But when pandemic occurred and schools were closed to curb the spread of viral infection , the EdTech companies provided a strong digital platform from where student and academicians can interact with each other using this technological handy application !

Millions of subscribers registered to these services and in the matter of one year , these organization become well-known name in the EdTech sector ! So , this is how business research create new markets to gain the market advantage !

24.3 Business Research for cross functional association :

Friends , whenever a new technology enters into the market , its applications are seen in many cross functional fields for necessary system upgrades to reduce cost , increase efficiency and provide better working environment for customer and staff !

Business research done for interlinking of multiple technologies comes under this type of research ! It can be explained by example of robotic surgery typically happens in multi-specialty hospital with the help of artificial intelligence !

Courtesy: CFV, PIxabay

Here how many technologies are working together ? Here advance health science is joining hands with industrial automation , artificial intelligence , machine learning , special electrical arrangement , radiology , digital sensing technology and such type of supporting and monitoring technologies from where the surgery can be done efficiently !

If you see modern air plane , how many cross functional technologies are linked inside ? If you see the example of dairy products , how many technologies are interlinked ? If you see the example of building construction , how many technologies are interlinked with each other ?

When you have to check such type of cross functional interlinking facilities , all field experts experiment with mixing of such technologies and thus the final results are produced as a new product !

But who decides such type of mixing feasibility ? The answer to this question is the lead researcher or main business leader ! Main business leader has excellent networking with cutting edge technology creator and they do attain such type of national – international exhibitions time to time ! When an appealing technology is presented in front of them , with their creative business thinking , they ask for its implementation feasibility inside their organization , when the solution provider analyses the business model , they try their technology in two -three phases and looking at the success of each phase , the subsequent decisions are taken ! The initial mixing and matching period of one year is important , when the mutually important variables match with each other , the traditional system get upgraded ! This is how every organization upgrade themselves with respect to changing business needs !

24.4 Business research for live issues :

Do you have knowledge of every field ? No ! Do you know what will happen tomorrow ? No ! Friends , this is what about business research for live issues !

You carry research of fields which are known to you ! Many people have carried out research in that field and hence there some kind of literature available for research !

Courtesy: Pali Graphicas, PIxabay

But what will you do , when no literature is available and you have to solve a live problem ? Friends , here the role of core research comes into play ! You have to study the live problem with detailed analysis . You have to check which theory can be explained for its understanding !

Then you have to carry out experiments to observe its typical behavior and when you are certain about its protective formulation , then you have to try that formulation on test species and if that results are correct , then only you can try it on human being !

Once such trails become successful , new chapter get added to research world and later many researchers put their efforts in its future discovery !

So, which are current live issues and what type of research is going on ? Currently business research is going on climate change ! Research is going on bio-technology and weather resistant crops ! Research is going on uses of artificial intelligence and it's all field applications !

Research is going on health sciences to treat brain related ailments !

When all such type of research will be done to deal with live problems , respective businesses will foster further to provide world class research product to people suffering from those problems !

In next step , we will see , research for humanity ! ✳✳✳

MULTIPLE-CHOICE QUESTIONS

1) Given a chance which market you would like to research for business ?

   A) A brand-new product market where there is no other competitor
   B) Saturated market
   C) Growing market
   D) Dynamic market

2) How many fields are interlinked when invention like mobile application software is developed to grow online business ?

   A) All fields where products and services can be made available by seller – software – buyer mechanism.
   B) Only software services
   C) All routine services of rural area
   D) All fast-paced services of urban area

STEP 25 : RESEARCH FOR HUMANITY

Image Courtesy: Open Clip Art , Pixabay.com

25.1 Introduction :

Friends ,

As we have seen from start , the research is started out of curiosity around this very nature ! How the things are set in the nature , how they are interrelated , how they behave under application of typical forces was the main motive behind some of the early inventions ! Once the basic research logic is understood by the scientists , the further level of research simplified that concept and it has resulted into full-fledged model of proven invention !

Research for humanity covers deeper aspects of research which will not concentrate on profit making deals or possession of extraneous powers to specific bodies , but it consists of researching things which can save humanity from crisis situations , which will provide better living conditions to humanity , which will allow humanity to have faith in each other and such type of social influential things !

In this discussion , we are going ahead with different type of research work whose main aim was revolving around humanity !

25.2 Human resources development research :

Earlier, when companies started their operations, it was fundamental boss-employee relationship ! In this model, boss used to tell work to employees or workmen and they used to complete that work ! Wages and salaries of that work was paid off and this used to continue for long time !

Courtesy: Canvas,Pixabay

The concept of human resource development is appeared when organizations felt the need to develop their talented manpower to tackle the current and future business need when business started growing beyond their regional boundaries and at a time when they were accustomed to work with diverse work cultures and ethnicity !

The mannerism, etiquettes, organizational behavior become important ! The need of learning foreign languages and setting things up in presentable manner become a norm for international office culture ! People were treated with decency and respect for fostering positive work culture across the organization !

So , the field of human resource development is researched for job role description , key result area determinants , performance appraisals and salary payment structures , employee lifecycle management , recruitment and compensation schemes , incentives and emoluments for accelerated performance , annuity age and retirement benefits , employee health care and insurance needs research and provision of partial or full medical cover , overseas deputation authority and method of payment of additional perks , hierarchy and organization structure to keep organizations moving and progressing , linking of financial activities to business momentum by using people's network !

The role of head of human resources managers is considered at par with role of managing director or vice president ! Decisions taken at corporate level were done with noting the effect on workforce as per opinion of HR head ! For resolving interpersonal and trade related conflicts , HR practices are frequently used to have fine dialogue and negotiation with the trade groups and employee unions ! So , human resource manager acted as single point contact from management side for settling various people issue to keep business moving and prospering ! So , if now you are able to see the growth of people policies in any organization , the result is because of extensive research done in the field of humanity and best working practices !

25.3 Research done in human psychology :

Understanding human psychology is fundamental research of any human resource development ! Until you understand any persons personal and professional background , how can you fulfill their todays and tomorrow's needs ?

To understand the human psychology in simple way ,let's take example of recruitment of two employees just passed from premier engineering institutes ! Mr. A is son of a farmer whose father's annual income is 3 Lakhs ! Mr. . B is son of a doctor whose annual income is 3 Crore ! Both are joining this organization at an impressive pay package of 30 Lakhs per annum ! Mr. A has to create many assets which are own house , own car , own jewelries and such type of basic human expectations ! Mr. B is well settled , he has everything which A has to own ! So, his whole salary is just regular addition to his family wealth and he can work on this package for long long time if the job is really satisfactory in long term !

Courtesy:Padrefilar, Pixabay

Mr. .A has to spend half of the salary for his household need and half of the salary towards payment of various EMI ! He has to wait for salary rise for one or two years , if it doesn't happen , he has to start some side business to raise his earnings ! He may think to switch the jobs and gain a salary hike if his financial needs are increasing !

So , naturally the psychology of person with more personal responsibilities keeps growing and the person with sound financial background can take enough time to take important decisions ! His mind remains stable and as a result he lives a quality life throughout his lifetime ! This is what expected by every human being to reduce the level of life struggles and raise the level of life comfort ! The research done in human psychology thus helps to gain the typical human insights and various experiments done on typical human behavior !

There are various human traits . Human can be introvert and they can be extrovert ! So , in this research , human thinking and the job requirement of that position will be researched to find out the best desired qualities in a person for that role !

If the role is expressive like an actor , communicator , political leader , introvert psychology will not work there ! If the role demands clinical perfection and attention to details , the person need to be introvert !

The surrounding in which human being has to work also affect psychology of that human being ! The person working with a creative music band is a jolly person who like to whistle and play soulful songs ! The person working with logistics is always run behind the consignments one after the another ! A person can be a teacher in the day and he can be a club cricketer in the evening ! So , based on the available surrounding , human psychology changes accordingly ! These changes are noted in psychology research done to understand humanity as whole !

25.4 Research done on customs and traditions :

Can you offer a supercool ice cream to person living in Jammu -Kashmir ? Can you offer coconut water to person living in Kerala ? Can you offer sugarcane juice to person living in rural areas of Maharashtra ? Can you offer lassi to person living in Punjab ?

The answers of all these question in ' No!' This is because all these people enjoy these stuffs in their own territory ! It's their tradition and some of their age-old customs !

Courtesy: OCA , Pixabay

So, the research done in understanding such remote localities help organizations to pitch their products as per the local need of the people ! Before you launch a typical product, it is necessary to know the local customs and traditions of that region to sell those products !

To make business easy and progressive, people advertise their product by keeping local costumes, local dialect, local artist into their mind ! This is because impact of local connection has important contribution to faith of the local people ! There is natural affection and hence it become easy to believe in the product when the local hero or artist promote it or talks about it !

If you know, in a particular region Diwali is celebrated for 10 days instead of 5 days as national average, your product sales and their consumption will double for this additional period of extra 5 days ! So, you will typically research which people celebrate Diwali for 10 days, what they purchase, how they spend money in those 10 days, how many guest come to their home, how many times they meet each other ! Such type of insights gives you ample space to launch your products which will meet their 10 days needs !

On the other hand , if your competitor is not aware of this special fact , your product sales will go on increasing and thus you will always remain ahead in your business dealings because of this research !

In another example , suppose in a hilly area , there are 5 months of cold waves ! People use warm jackets often and they prefer to keep one extra jacket compulsory in case there is further lowering of temperature ! A firm manufacturing warm jackets understand this typical habit of local people and they launch a scheme of two jackets with 25% discount !

More people buy this pair of warm jackets and earn the descent discount of 25 % which save almost 500 Rs as compared to other producers ! The typical local area has population of around 10 lakhs and almost everyone buys such type of warm jackets for their dear ones ! Now , just think , a simple traditional habits & customs knowledge give that much benefit to warm jacket producer in comparison to its other competitor ! Again, this research is providing financial benefit to humanity ! After every two-year people buy such warm jackets and thus business get its regular customers !

25.5 Research done on lifestyle concerns :

People like to upgrade their lifestyle and hence this field of research is developing constantly ! There may be upgrade in home selection , clothes selection , jewelry selection , car selection , this habit changes continuously and accordingly makes has to design and offer their products ! This is about luxurious lifestyle !

Life of common man also upgrade but the financial investment is comparatively lower ! Researchers has to understand what type of two-wheeler common man use for himself , which type of clothe he frequently buys , what are his food habits and how many times he prefers to take lunch or dinner outside his home , these research points create some of the interesting products and sales of such products keep improving !

In next step , we will spread light on research done for agriculture ! ✲✲✲

MULTIPLE-CHOICE QUESTIONS

1) **Which of the following is most challenging human resource development task ?**

A) Hiring 1000 Graduate trainee engineers from NAAC A+ certified institute.

B) Retaining 50 star performers who are open to lucrative careers outside the organization.

C) Implementing the cost reduction projects across the whole organization as per managing directors action plan.

D) Setting up a new venture in foreign territory, understanding detailed lawful framework, receiving various statutory and commercial permission and starting business by hiring people, buying material and machines and making customer comfortable with the new venture !

STEP 26 : RESEARCH FOR AGRICULTURE

Image Courtesy: Open Clip Art , Pixabay.com

26.1 Introduction :

Friends ,

Food is the basic need of human being ! Survival of human being without proper food in next to impossible ! Human being can live their life by just drinking water but the energy required for daily work is received only through nutrients contained in food !

Earlier humans use to do hunting as well as they used to eat fruits and leaves of some eatable trees ! Hunting was used to collect food for daily living ! Like in a jungle big animal kills small animal and make his living out of it , man was also doing the same when he was living in jungles !

As man started living in civil society away from jungles , the invention of agriculture is occurred ! Man has understood , there are some type of food grains available in this nature which can be grown systematically in local farms , the water supply of rain will ensure the growth of these crops and when the growth will be completed , the final product can be used for human food needs !

So, man started experimenting different food grains in different climate conditions available ! Climate all over the world is not same and it has huge variations ! So , he observed that most favorable crops will grow fine in such a climate and with proper care of plants better yield can be obtained !

As far as India is concerned , it got good quality soil in which many foods related crops can be sufficiently grown ! India has got good climate to grown many fruits and vegetables as well as high yield commercial crops !

So , farmers tried different crops in available environment and observed its growth and taste ! For example , Assam tried tea plantation and it received excellent results ! Same thing is done by Tamil Nadu and Kerala ! Maharashtra tried sugarcane and got good result ! Uttar Pradesh tried for wheat production and got excellent results ! Punjab tried for wheat and they also got good result ! Jammu Kashmir tried for apple and got good result ! This way every state got some typical crops which become the identity of that region ! All people ,noting that type of climate planted those seeds and earn their living by its huge production !

These goods were traded in market with proper selling prices and from one village to other village trade is started ! As the commute options are developed , this trade reached up to district level ! When further road construction is done , interdistrict trade is started and which is followed by national and international trade !

This is how the work done in the field to meet personal food needs become an opportunity of trade across the nations worldwide and this given rise to agriculture-oriented economy !

In this discussion , we are going to spread light on different avenues of research done in the agriculture sector !

26.2 Research done for Soil Quality :

Soil is the backbone of every crop system ! It is the strata or foundation on which crop grows ! Crop receives many nutrients for their growth from nature of soil available !

So , research done in soil helps team of scientist to recommend the type of crops which can be grown satisfactorily ! Some type of soil has good amount of water holding capacity which is necessary to supply required water for crop growth ! Some type of soil retains very less water and this quality is also helpful for some crops which do not require more water ! Some type of soil has lots of mineral content and hence the crops which are grown in such type of soil are mineral rich !

If the surface of earth and geography of India is concerned , India is blessed with number of mountains and rivers ! The rivers originate mostly in these mountains and they flow towards platue region until they merge with sea ! In this journey , the transfer water from mountain to platue region and along with that they carry rich quality soil from these mountains ! When this soil settles in platue region , the land become rich of these minerals and thus the crops grown in these platue area gives good growth to those crops ! These crops in turn provide required nutrients to people who consume it ! Some type of soil has iron content above exceptional level, so not many crops can grow in such soil base!

Some type of soil is extremely sticky and specific crops are grown in such soil! The agricultural universities study such type of soil types and recommend the best crop system which can be grown in such soil base!

Courtesy: Bored, Pixabay

26.3 Research done for seeds and their usage :

This is also one of the most important agricultural research fields! Here detailed study of crop species in done in laboratories and the growth cycle of crops is observed! Various experiments done to increase the yield and reduce cycle time! In some experiments good success is achieved while in some experiments undesired results are achieved! Scientists take note of both things and recommend only useful seeds to the farmer community!

The ideal seed need to be clean ; it must not have any type of internal and external defects , it should be free from any type of infection from insects , bees and other microorganism present in that climate ! As it later gets converted into many crop diseases which halts crop growth and also reduce total yield in that farm !

Some seeds have good fruit generation capacity ! When you plant one seed , the fruit of that fully grown seed plant gives hundred seeds of that type ! So , such type of high output seed is desired by every farmer as it gives more fruits in less area which is beneficial for small farmers which have just one or two acres of land !

Seeds also need to be easily available for commercial farming in hundreds of acres at one time of particular season ! If the seeds are not available by that time , farming season may go without any production !

So , agricultural institutes spread across the national region try to produce enough seed capital with which every farmer can buy those seed bags at suitable price and can sow them in their field !

The next step of seed research is making of plants and baby tree which can be planted into open land ! This space is generally known as nursery ! Here, small plants are prepared and sold to farmers, farmers receive them and plant in their fields ! Lakhs of plants are sold by every nursery to farmers and farmers plant them in their fields !

Courtesy: GDJ, Pixabay

So, food grains and tree plants are two types of input which are fed to soil and then with their all-season care, the output is taken by farmers !

In the age of biotechnology, bio technologists are developing crops of high yield, low growth duration and more nutritional values ! Weather resistant plants are also under research in premier research institutes of the nation !

26.4 Research done in climate change :

Climate required for agriculture is an important parameter ! Climate changes are unpredictable and they are mostly indicative . If someone consider the typical raining pattern , no one can accurately talk about the possibility of rain ! It's uncertain and hence more research is being done in the weather forecasting aspect !

Now people have launched satellites around earth radius so that they can predict the weather conditions and possibility of rain ! In summer time also , prediction of maximum and minimum temperature can be done and same thing for fog intensity in winter season can be predicted ! All these factors affect typical crop growth and hence research done in climate changes will benefit farmers community to huge extent !

Another research done in the climate change section involves study of growth of various microorganism and its effect on crop systems ! People use fertilizers and insecticides for good growth of crops ! The effect of chemical fertilizers and insecticides are not good for human health if they are used in intolerable proportion ! People are also researching the possibility of developing natural farming system where bio fertilizers or organic fertilizers can be used for crop growth !

Looking at the global climate change , the effect of sudden rain , low raining , high amount of summer temperature and very very low winter temperature are affecting crop growth adversely and for this reason efforts are done in the research laboratories to develop weather resistant crops that can tolerate important climate changes !

26.5 Research done in irrigation schemes :

The most important need of farming is the supply of water ! Traditionally , water needs are fulfilled by natural and man-made water reservoirs ! If the farm in located in the vicinity of big water resource such as river or lake , the ground water level remains high and it supply water to the roots of crops !

Based on water availability , the crops are classified as seasonal crops and all-season crops ! All season crops get ample amount of water from various water resources located in their vicinity ! Or farmers located near water resources prefer to take multiple crops as per various seasons as water is available in ample amount ! In contrast , farmers located in low water availability rely of water supply from natural rain and its proportion ! They do have well and small lakes in their fields but the water availability is always a challenging condition !

So , they prefer to take crops that require less water and whatever crops grow in that environment are sold to people !

Courtesy: GDJ, Pixabay

Looking at the need of water , research is done to supply water through sprinklers and drop irrigation method ! In this method , optimum water is used and wastage is avoided ! In mechanical farming , field equipment's are used to distribute water as per required needs and some proportion of water is also recycled for other tress and plant nearby the actual farm fields !

Diverting natural flow of water to farm fields is also researched by carrying out water development plans . In such plans water available in dams and water surplus regions is moved through canals to nearby agricultural zones !

In next step , we will see , research done for good life ! ✷✷✷

MULTIPLE-CHOICE QUESTIONS

1) **Which of the following agricultural research field is facing serious impact to its existence because of heavy rainfall and global climate change ? Incidences of frequent rainfall , rising proportion of reduction in natural greenery , increasing pollution levels are adversely affecting to this reseach field !**

A)   Research For Soil
B)   Research For Seeds
C)   Research For Irrigation
D)   Research For Crop Insurance

STEP 27 : RESEARCH FOR GOOD LIFE

" These are special knowledge fields and it has sensible connection to social stability and social progress ! In fact, many political and social leaders use the output of these research fields to control, manage and lead this society with grace and affection ! Hence this type of research work is basically known as research for good life ! "

Image Courtesy: Open Clip Art , Pixabay.com

27.1 Introduction :

Friends ,

In the concluding section of this book , we are moving our attention towards some of the research fields which are beyond mere scientific studies and inventions ! These research fields may be called as stories beyond invention ! These fields have deep meaning of overall human Lifecyle and making this lifecycle smooth and easy for everyone ! These are special knowledge fields and it has sensible connection to social stability and social progress ! In fact, many political and social leaders use the output of these research fields to control , manage and lead this society with grace and affection ! Hence this type of research work is basically known as research for good life ! Every government is ought to provide quality life to their civilians as well as their in-force members ! The job of every politician is to keep this human society free from major conflicts , engage in constructive work that builds your nation and spread the vibes of co-operation with like-minded countries to become part of a globally respected human civilization ! So , let's move to the typical research fields which carry out their work to live a good life !

27.2 Religious Research :

When people are fascinated with questions like who I am , why I am present on this planet and what exactly I have to do in my life , they swiftly move to concurrent literature which gives this information and knowledge about ones 'Dharma' or religious duty of righteousness ! Duty of supporting what is right in any situation ! This ultimately avoids wrong from happening and hence avoids all chances which generates moments of sorrows from such acts, behaviors and incidences !

In this world , there are many religions are present and every religion firmly believes in preachings given by their divine prophet during

Courtesy: GDJ, Pixabay

their lifetime and these preachings are further cherished by their followers in the form of literary treasure known as Religious Literature !

The life of many prophets is an inspirational engagement for their followers and believers ! These prophets always stood behind the righteousness even though they have to suffer major stressful incidences from that time's society !

While facing pain from society , they always communicated the power of prayers for healing someone's deep pain and they also guided the society to believe in the values of humanity which are nothing but freedom , brotherhood and equality ! When these values are imbibed in society , automatically the feelings of love, truth and service get generated and it give rise to a selfless and happy society ! This was what the duty of every prophet performed sacrificing their own comforts for this huge human as well as non -human society !

In science , we study when chemical A mixes with chemical B , it gives rise to reaction products C & D ! Here , C can be present in one form and D can be present in different form ! So , noting the reaction mechanism , scientists formulate this reaction and record it on paper ! The interested people , refer this formula and carry out the same experiment at their end ! When they receive same result , it become a moment of knowledge transfer ! Later on, this knowledge is used for meeting human needs of different type and when these products are sold for a price , the trade relation get established ! This trade relation goes on for generations to generations till that principle is present in this nature ! With change of time , there may be little bit variation in reaction outcome because of reduced or enhanced reactivity ,but the common reaction output never ceases !

This is because , this is what the nature is ! Nature has some set principles which are not firm and that is why there is stability in nature , because of natural principles , there is possibility of human and plant life , because of these principles one can understand the nature's season cycle where people experience diversity of nature ! These principles are nothing but five major powers present in this very nature ! These are Fire , water , Sky , Wind & Soil !

Because of the proper engagement of these five mega powers , the nature behaves constructively for this society and when there is any type of imbalance , the radical seasonal changes are observed from time to time ! Such cycles of changes are occurred since past times and the society which is living currently has gone through all such type of radical natural changes !

If someone carry out this study of human evolution , they will notice these natural changes took place million years ago !

So , moving again towards religious literature , prophets mentioned these natural phenomena because of their detailed study of this nature since ancient times ! As far as Hindu religion is concerned , the knowledge of this nature is expansively available in *Vedas* and each *Veda* is dedicated for special concern for this humanity and civilized world !

Not only *Vedas* but many saints , prophets , monks and social servants written down their learnings in the form of books and various types of literature ! People used to write such type of important preachings on the walls , so that any people passing from that wall will possibly read that preaching and make himself aware about the reality of life! Because of this, people used to follow some typical civilization rules and in case someone breaks that rule, he has to face the relevant consequences in the form of penalty or hardships of different nature!

Courtesy : CFV, Pixabay

So, the religious researchers study these types of ancient literature and find their relevance in today's society! If they notice some society issue and if they know this literature has given solution for this society concern, they document the problem and solution in their research work and they carry out social experiments where those type of problems are visible! They implement the solution given in the literature or they develop their new solution by referring to that literature and observe its impact! So, if impact is positive, they record the results and present this research work to their research guide! If the work is commendable, they get the appreciation, if the work needs more back up in the form of experimental data or additional literature support, they have to again refer to another ancient literature !

The society get mature over period of time with different natural and man-made experiences ! The generations living in 2024 are not aware about how the wars are fought in 1600 with the help of army and horse power ! But it is mentioned in the periodical literature which is known as history of a typical nation ! This is not written today but written either when it was happening or just after it is happened ! That times government preserved this literature to understand the political relations, rivalries and mutual co-operation so that future generations will take care of such painful experiences ! This is what the intention behind research for good life ! People of earlier generations has told to future generations to stay away from conflicts and wars, as they are not good for your very existence of your life on earth ! Instead ,choose peace, love ,co-operation because of which you will also survive and your friend will also survive and you both will live happily !

So , all forms of religious and historical literature available has said the same fact of how to preserve human values , how to protect this nature and how to maintain peace in the society ! When researchers carry out work of such type and write down their study result for living good life in today's era , the research work get people's appreciation and people start using that approach to live life in today's age ! This is also called as living life in *New Normal* mode ! Any type of abnormality is not good for this nature and human existence and hence this religious preaching will always act as the first and last solution over various social issues !

27.3 Life literature research :

There are two ways of living your life , either refer what your religion has recommended for happy and descent life or live your life as you like without considering any society conditions and norms !

This is the what main pain point of living your life as you like ! This is not allowed by society ! Society means huge number of people , it means majority ! If you are doing something against majority , naturally opposing force will get activate and depending upon the intensity of deed ,society will put different restriction on that person or group ! The society will put that group differently and will avoid any type of relation with such people or groups !

This is where the life literature finds its place ! People who realized their life differently , they opposed to many society customs and rituals and they proved in their own way that such type of things generally do not matter to basic human life which has deeper meaning of serving each other selflessly !

Courtesy: GDJ, Pixabay

So , if you research on life philosophies of various authors , poets , social scientists, political and sport leaders , you will notice the generous nature and life understandings behind their work ! In most of the Marathi poems and literature , poets best described the way of living life happily and progressively .

This is about life literature ! It is expressed by creative individuals may be a hobby , maybe as a social need of that time or may be for some commercial benefits ! But this life literature is also researched by many researchers to lead a happy and content life ahead !

It's well-known fact that reading literature enhances critical thinking and decision-making ability of any person ! So , when researchers research on life literature , their life understanding become broader and hence they can take many life decisions without any fear or worry because they know what is going to happen if this decision taken at that time !

So, every researcher of good life first benefit himself and then he shares his experience to community in which he is living ! Those who adopt that learning also enjoy its benefits and thus a society of high and rich life experiences get created ! This is one of the most useful benefits of research done for good life !

In next step , we will look into research done for liberation ! Another story beyond invention ! ✲✲✲

MULTIPLE-CHOICE QUESTIONS

1) **When the feeling of self realization is experienced by a person doing religious research ?**

A) After visit to spiritual places .
B) After reading spiritual literature .
C) After doing service of poors .
D) All of the above .

2) **Which of the following spiritual literature quotes about the life values , life purpose , life goals , life struggles ?**

A) Poetry composed by Saints .
B) Life Experiences shared by religious masters after prolonged spiritual practice .
C) Compositions compiled from folk -stories.
D) All of the above .

STEP 28 : RESEARCH FOR LIBERATION

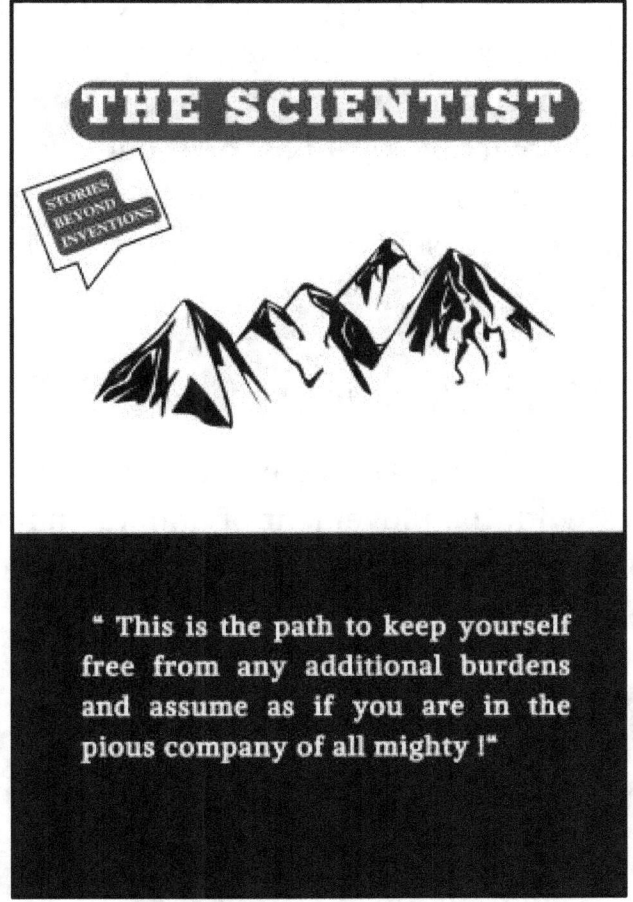

Image Courtesy: Open Clip Art , Pixabay.com

28.1 Introduction :

Friends ,

In last step , we have seen ,research for good life , in this step , we are moving towards broader terms of life understanding in the form of research for liberation ! It will be interesting to note that how the quest of understanding this nature around us literally comes so close to knowing yourself !

28.2 Research on self-consciousness :

A devotional wave was generated in the west with the movement of understanding self-consciousness ! People having all sort of material comforts , facilities , support staff, heavy financial strength were still not as happy as they supposed to be !

This point raised the debate on awareness that there is something beyond the science and it is nothing but the kingdom of supreme spirituality ! Yes , spirituality attracted these super rich and most advanced human being to find the true essence of life !

When they entered in this kingdom under the able guidance of the Guides of this kingdom , there overall lifestyle changed to unimaginable super simplification ! In some words ,one can say

Courtesy: CFV, Pixabay

, the removed lots of baggage from there mind and they started feeling light and without any worries !

This was the path of self-consciousness and moving towards liberation ! This path of understanding the birth and death cycle and understanding what is to be done in this lifetime in between ! This is the path which ask for total surrender to the almighty and then he will take care of your consciousness with his magnificent love , care and bond of permanent attachment ! This is the path to keep yourself free from any additional burdens and assume as if you are in the pious company of all mighty !

The service of poor's , cows and age-old seniors is suggested as means of finding ultimate solace to the entrants ! Service to these classes is nothing but service to almighty ! This thought process made these entrants extremely down to earth and soon they started realizing the true joy of total surrender !

There are multiple preachings on the path of liberation ! People were guided by these saintly preachings through *pravachana* which is also known as speech delivered with spiritual experiences of high impact ! Through these speeches , the saintly figures literally demonstrated the path of ultimate liberation to the disciples' and thus it started their understanding of knowing themselves !

Every stanza , every line and every incidence of his divine grace was explained in most simple and prolific manner to disciples with which they got overwhelmed and started singing and dancing in the glory of almighty !

Food habits of disciples are also changed once they entered into this pious kingdom of love and faith ! They become vegetarian and observed themselves free from many diseases that initiate because of high calorie food habits ! The disciples' started services like cleaning , washing , decorating places and thus they observe , how nice it feels when one does such kind of simple task under the grace of almighty ! The day of disciples were closed by the various songs dedicated to almighty's absolute bliss !

People entering such kingdom literally forgot their existence in materialistic world which is full of agony , anguish , competition , anger and other such type of *Shad ripus* – six enemies of mankind ! They entered into the company of *Satvik* environment , pure environment !

For sake of easy reference , they have given different names and these names were nothing but calling of almighty's name ! So , all the disciples' present there were bearing any name of almighty and thus throughout the day there were chanting almighty's name even they are calling each other for simple reasons ! Chanting name of almighty is nothing but connecting your soul to his soul ! And if this type of connection is happening on daily basis and with frequent spells of chanting , this sound resonates with the ultimate of path of liberation which merge with his pious soul !

The person entering this kingdom of faith, respect and love was free from typical elder years sufferings ! This kingdom treats everyone as they are ! This kingdom accepts the lifelong service of the disciples' which they have done for their families as a part of responsibility and duty of their own ! This kingdom heals their internal pain and make them blissful by sharing the grace of almighty !

Courtesy: OCA, Pixabay

The kingdom of liberation offers food, shelter and rest with provision of living simple, sober and gentle life ! In this life, you talk good with each other, you do good work only, you read only good literature and you think only good ! Whatever looks impossible you to achieve, you surrender the same to almighty and keep doing your duty sincerely, to your pleasant surprise, on one fine day, that impossible work becomes possible ! When this moment of realization clicks , you become the supreme devotee of his divineness and you surrender again and again to his blissful and joyous atmosphere and this makes you extremely polite, humble and kind !

The rise of good human emotions like apathy, kindness, lion heartedness, purity, openness comes when you spend valuable time in the company of his divineness ! So, this path of liberation is nothing but research carried by you on your own life ! Once your research completes, you start sharing your experiences with due approval from your spiritual guide and the next generation of guide -disciples' get started ! So, new entrants enter into the kingdom of almighty and they also walk the path of total liberation and attaining heavenly bliss in the presence of his divine grace !

There are typical universities and spiritual centers which award such type of holistic upscaling of their disciples' and award them relevant designation as per their spiritual designation system ! The spiritual powers experienced by any disciple is nothing but the huge amount of work they do for this society, the huge amount of support they provide to diseased and poor, huge amount of contribution they do for education with which people can become self sufficient ! This is selfless service done without any expectation of money and reward ! The disciples' do this service because they have totally surrendered to almighty and under his divine grace, they feel extremely inspired to do this service ! They experience the difference between earlier satisfaction of material pleasures and comforts and they also know the quality of this devotional service to needy !

Many disciples' entering this kingdom prefer to live as bachelor or as *sanyasi* ! One who doesn't have any physical & emotional attachment to any opposite gender ! The path of liberation of such disciples' starts from having natural birth from their parents , completing formal education , feeling of self-realization in early stage of life and then entering into the kingdom of almighty for rest of life ! They totally devote their life for service to weaker sections of the society to make them stronger and thus reducing the gap between rich and poor !

Their almighty has everything ! He holds supreme knowledge , he is supreme rich , he can support any number of lives , he holds all powers of survival yet he is in quest of real devotees who totally surrender to himself , he waits for those devotees who do good deeds , he take care of those devotees who chant his name , chanting his name just gives him the joy of being with real devotee, his relation with real devotee is not like a master and slave but it is like a friend meeting other friend ! He is helpful , he is kind and he is affectionate ! One cannot overlook his divine presence in the form of positive energy around his symbolic places such as temple and *ashram* !

He is present all where and you just need to remember him from your clean and clear intentions ! When such type of real devotee forgets his real existence , he just starts overjoying with his grace and affection ! For normal people who are living a materialistic life , this may look as little bit different but the person who has surrendered himself to almighty is just like an innocent kid who is enjoying the company of his affectionate mother !

This is what the meaning of liberation ! The physical death of such person also happens as per normal human being but the habit of chanting his divine grace make that final path quite easy because the energy of chanting make him internally strong to bear the pain of aged body ! This is where being vegetarian helps them lot ! They have normal physical body and with the practice of *Yogik kriya* they attain a strong body in their elder age also !

Courtesy: GDJ, Pixabay

Samadhi is the process of experiencing the liberation and it is the final goal of many disciples' entering this kingdom ! There are types of *samadhi* and the purpose of attaining state of *samadhi* is merging with eternal bliss forever ! There can be symbolic gestures when some saintly person takes *samadhi* ! That place gives you peace and positive energy ! That place also gives the experience of purity ! That place gives your courage and uplift your morale to fight life issues ! That place makes you comfortable in any type of tough life situations ! This is because , this saintly person has devoted whole of their lifetime for serving others selflessly and hence they become master of understanding materialistic life's pains , its resolution methods and the ultimate receipt for them which is the bliss of his divine company which is felt time to time while doing something better for others !

So , if you notice the path of liberation , it is nothing but moving towards less *bhogas* – habits of materialistic consumption and accepting the habits of *yogas* – experiencing his divine grace by following the path of yogic process known as *ashtangyog* , the process of self consciousness where a person train himself under the company of able yogic guide and attains a state of *samadhi* – liberation !

Yogic people always consume less , they keep fast and they eat just sufficient for living ! They do only good deeds and refrain themselves from any kind of negative emotions ! They believe in almighty and keep chanting his name as a mean of knowing themselves and their connection with almighty ! So , this was the little bit about the story beyond invention covered through research on liberation !

In next step , we will look into research on values ! ✷✷✷

MULTIPLE-CHOICE QUESTIONS

1) Many academic, spiritual and social organizations starts prayers before commencing their work. What are the benefits of prayers for any scientific researcher who does not believe in powers of prayers?

A) Prayers are positivity creators and words used in prayers are aligned towards overall goodness of everyone!

B) If a person does not believe in God but indirectly listen prayers when it is started nearby his residence or workplace, it relaxes their mind because of humming united public voice which is symbol of solidarity in testing times!

C) Researchers listen to prayers and keep doing their work as they like to do with neutral approach! Prayers starts and stops after serving for goodness.

D) All of the above!

STEP 29 : RESEARCH FOR VALUES

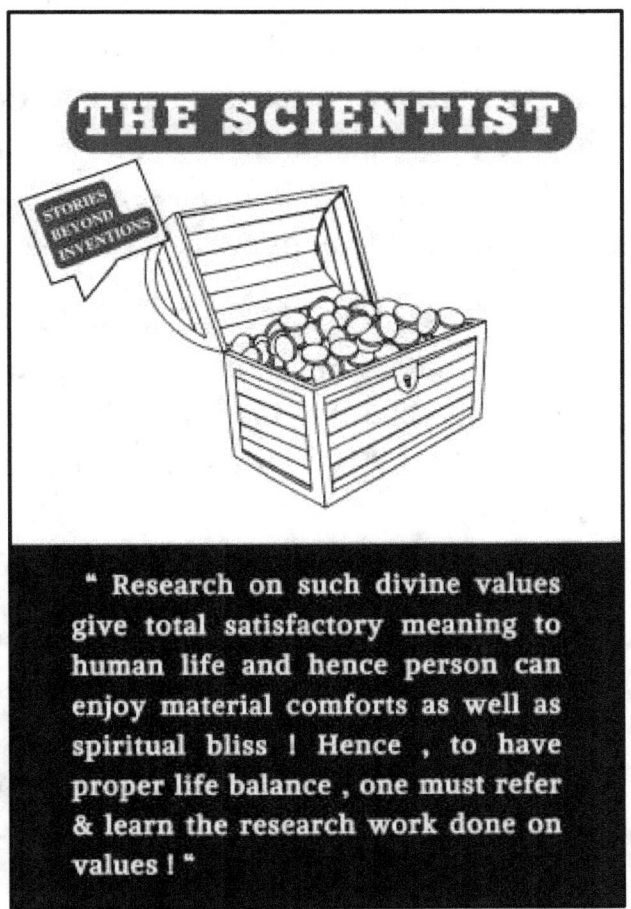

"Research on such divine values give total satisfactory meaning to human life and hence person can enjoy material comforts as well as spiritual bliss ! Hence, to have proper life balance, one must refer & learn the research work done on values ! "

Image Courtesy: Clker Free Vector , Pixabay.com

29.1 Introduction :

Friends ,

In last chapter , we have seen research on ultimate experience of liberation through the process of *Ashatangyog* ! The last perfection of this research is *samadhi* ! Unison with the universal soul in his blissful presence for utter joy and unimaginable happiness !

Now , we are moving towards , the research on values ! Human values ! You may think , why this chapter is listed after the end state of liberation ? But this is the specialty of this discussion !

When the great personalities live their live in this world and do the social work for which they were incarnated as per spiritual beliefs , what remains after their liberation is the blessings of proven values , which are extremely valuable to lead a rewarding and happy life !

Research on such divine values give total satisfactory meaning to human life and hence person can enjoy material comforts as well as spiritual bliss ! Hence , to have proper life balance , one must refer & learn the research work done on values !

29.2 Research on Personal Values :

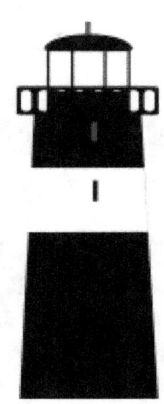

Values are identity of a person which are developed after several years of life experiences ! Development of values has close connection with person's basic upbringing , his family background , his surrounding , his schooling , his friend circle , his college education , his college life , his job experience , his job progress track record , his personal family life after marriage and how he inculcates these values in his children's , his children's life and his role to shape their life with good values , this is how personal values are researched to learn more about any person !

Courtesy: CFV, Pixabay

This learning of values gives an observer a chance to find out avenues with which this person can justify roles given to him satisfactorily ! This research is done by many human research firms before appointing a crucial leadership and supporting roles !

This is because , more personal values a person belongs to his kitty , more promising his career decisions will be ! Recruiters need people at leadership role who have never say die attitude , who are humble for continuous improvement , who are straight forward in knowledge sharing , who are considerate to understand their colleagues , who are practical enough to bear the rise and fall of business situations , who are jolly enough to spread the happiness and positivity in their work environment and above all ,they need to be frank to admit what is wrong is wrong but it can be corrected with good correction solutions !

To get a person of such caliber , researchers look for honesty , transparency , faithfulness , solidarity , toughness of character , smartness of presentation , quicker decision-making , broad-minded thinking and collective participation in group moments !

So, how a person's personal value system is ascertained before his or her appointment ! One of the proven ways is interviewing the person with open ended and closed ended questions and check how frank and true his answers are to the resume or bio-data he has presented ! The one page of bio data consists of ten years of practical experience hence at senior levels, they need to mention only critical achievements which stand out and which are special ! In interview, just the details of these highlights are asked to make a final decision of acceptance ! The bio data itself provide detailed track record of that person and for the interviewer, its daily job to check hundreds of bio-data before shortlisting a person for interview ! So, they easily recognize, what type of efforts this person may have taken to reach to that highest position. This is impossible without having a strong personal value system which keeps this person moving on and on !

Now recruiters come researchers has dual role to play before selecting a dynamic leader ! They have to understand need of current market need, they have to check current system performance gaps and then they have to decide what type of leadership will make this situation progressive ! If business is running in fine condition , they need a leader who can exponentially raise the business momentum ! If the business is trapped in difficult financial conditions , they need a person who can swiftly change this scenario with powers available with him ! If the business is just launched , they need a leader who has all round business experience and who like to coach youngsters !

Now , if the candidate is smart enough to understand the company's history and current condition , he will suitably respond to answers ! If the candidate has not done enough homework before appearing for the interview , he will show the personality gap and hence the recruiters will not think for his candidature !

Hence attention to details and planning with preparation is again perceived as an important personal value !

The development of personal values take place by practicing them and not studying them ! If you are helpful , you always help to everyone by sparing enough time from your work ! If you are kind , you always forgive people who make errors and mistakes and you try to remove the cause of error ! If you are a happy person , you always share jokes , funny things so that other people also share the same and keep atmosphere lively and energetic ! If you are cool person , you never get angry and you always think with rational approach ! If you are humble , you gently try to complete all work as fast as possible and you move to new work without wasting time !

So , when such kind of candidate is selected , he contributes to his organization by adding these values through his work methods ! He inspires his team to improve the current conditions and reach a platform which indicates graph of seamless progress ! When this stage is achieved , then only you become successful regularly !

When such person works in the organizational atmosphere and leaves the work after his super annuation , he has upgraded the work culture to best possible level ! During his stint as a leader , he has developed a team who is strong , caring , experimental , generous and fun loving ! By this time ,he has significantly contributed to organization to make a path breaking turnover that ranges from rags to reaches ! So , when such person leaves the organization , he leaves an ultimate legacy as inspiration for his successor ! The successor , again , a person with exceptional personal values consolidates the progress which his predecessor has done before him ! This is how the people are appointed and given a chance to build robust and performing organizations that values personal values to considerable proportion in their business activism ! This is why human resource is also called as Human Capital ! It acts as an asset for your business !

29.3 Research on Business Values :

Courtesy: Tanrika, Pixabay

Business is done by organizations and organization are built by the people using various resources ! First you built the constitution of your business plan and then you build the workspace where you will start your work with recruiting right type of people for specific role !

There are three to four layers in any organization and for each layer you don't require same skills ! Every layer requires different skills to carry out their daily work and hence recruiters recruit people with different skills but almost same personal values ! This is most important selection criteria for recruiting any type of role !

For example , for selecting a trainee apprentice , you will see his academic track record , his hobbies and his project work ! You will also see ; can this person perform well so that he can continue with the organization ! If so , you ask such type of questions . If you are not sure about his personal values , you give him chance of apprenticeship and see whether he demonstrate those required personal values in that training period . If he works well , you offer his permanent role in the organization . If he is not meeting expected performance criteria , you again give him six-month extension , then he also realizes the required improvement points and after attaining that level of performance , he is continued in the service !

This probation period is important for every layer ! More seasoned the person is , more easily he gels with the existing culture ! During the switch over period , new person has to leave many work practices of earlier organizations at their place and he has to start working as per current organizations work culture ! This point he has to take care before accepting his role by doing proper company and job role research ! After joining the organization , his role become important and he has to deliver for required business targets ! So , wise people always choose candidate which will be a best fit to organization culture and a right candidate also prefer to join an organization which has same work culture as per his earlier organization ! This makes company switch over a mutually rewarding task !

So , business values require strategic orientation ability , mission planning ability , vision setting and achieving capability , financial transparency and openness for change management , risk taking potential and understanding of market inclinations ! Business survivals skills and introduction of culture of innovation for the organization ! This research is continuously done and many business schools and institutes offer such type of academic upgradation courses for working professionals ! Academics always study the market trends and develop educational material so that many working professional upgrade their knowledge and skill to stay relevant to market needs !

29.4 Research on social values :

When you are done with personal values and business values , social values research starts with the process of socialization ! Which includes sharing things with each other , raising voice against injustice , providing fair treatment to everyone , enjoying each other's success when it is attained by fair means and providing support during testing moments ! The whole aim of social value

Courtesy: Paligraphicas, Pixabay

research lies with everyone is part of big, strong and affectionate family and everyone has equal rights as per the constitution of that nation !

In next step, we will conclude this book with research for advancement !   ✹✹✹

MULTIPLE-CHOICE QUESTIONS

1) Values that drives typical corporate governance because of which people exchanges their knowledge and fullfill their typical technical needs of their customers comes under which classification of values ?

A) Personal Values
B) Business Values
C) Social Values
D) Spiritual Values

STEP 30 : RESEARCH FOR ADVANCEMENT

Image Courtesy: GDJ , Pixabay.com

30.1 Introduction :

Friends ,

In the concluding chapter of this book , we are moving towards research for advancement ! Every research is meant for advancement to live better , reduce one's painful life and utilize the resources present in the nature for well-being of everyone ! Why nature has offered so many plants , fruits ,minerals and other offering ? Man has noted these resources and developed ways to utilize them for his materialistic consumption to full fill the basic need of food, clothe and shelter ! So, let's move towards various ways of future advancements .

30.2 Advancement in finding new materials :

This is one of the important research aspects ! New materials need to be invented for rising material demand ! After few days , material which is available on earth is going to be reduced and one fine day ,the natural stock will be completely over ! By that time scientists and researchers has to find a best alternative to that material so that people can use it for meeting their daily needs ! Material science and metallurgy is carrying out such type of research to develop new alloys of high strength and high applications ! Development of nano technology is considered as futuristic step of this material development ! Generation of ethanol from the sugarcane factory is also another method of finding alternative to other fuels ! The experiments of bio diesel are also carried out by researchers so that additional fuel options for mobility can be discovered ! The latest discovery of lithium batteries for charging electric vehicle will add more value to the segment of electric vehicle ! Gold and other precious metal are reducing is stock but new mines of gold are also discovered in some of the states and nation !

The research equipment's used to find new material are also advanced and with the help of latest technology finding a source of new material become more certain ! The machines like PMI

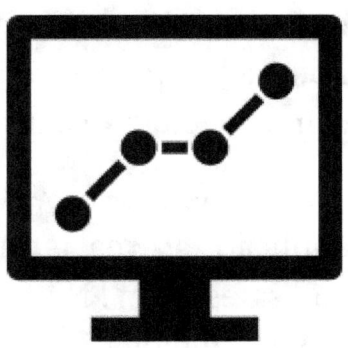

Courtesy: Tranquangkhai, Pixabay

– Particular material identifier also recognizes the elemental composition present in alloys and metals within fraction of minute ! So , such type of progressive developments is also carried out in material advancement research work to meet future needs !

30.3 Process advancements :

Process advancements including two approaches ! In one approach , you modify current process to increase the accuracy , speed and quality of same task ! In this approach , you make that process simple by inclusion of latest technology !

In another approach , you do the same task but by using different process ! Here also accuracy , quality , safety of products or services is improved than earlier processing !

When a new technology emerges , its application area is tested by design engineers and sales team ! Noting the critical domain area , this technology is offered to respective domain owners ! On finding its business viability , this technology is adopted and then the previous processes are upgraded to new technology !

30-40 years back, computers were rarely present in the office work in India ! They were present where computer related work is going on else other fields were doing their official work with the help of basic office documentation ! The inclusion of computers in official work changed the previous processes and offices adopted new ways of doing official work which is more accurate, fast and error free !

Courtesy: OCA, Pixabay

Then come the internet and any type of document transfer become easy ! Within a fraction of second anyone can send any document to any country ! This revolutionized office work and made communication super easy ! After this, smart phones are arrived and now office work can be made available of smart phones with the help of dedicated mobile applications !

So , the inclusion of latest technology has changed earlier processes and this advancement made work easy to do and people can enjoy what they are doing because of its ease of doing !

Another example of process advancement can be given for carrying out heart surgeries ! There was a time when heart patients were considered as most serious patients because of there was little possibilities of carrying out heart surgery in India ! Only medicines were ray of hope and hence people used to be ultra cautious about heart related diseases ! Then come the technique of bypass surgery and people got a chance to operate themselves under the keen and renowned surgeons in this super specialty ! As research advanced , the pre assessment of heart condition with the help of 2D echo system made decision making easy for doctors ! Unless there is need for open heart surgery , doctors advised preventive medication to treat their patients ! The technology is further improved and now artificial intelligence is also being used to find out the chances of surgery based on condition of heart ! Data of health can be easily shared with any doctor serving in the world and his second opinion can be taken before operation ! So, these process advancements are benefitting the heart patients !

So, every field and every researcher are trying to optimize their processes to provide enhanced output ! Enhanced quality levels ! Enhanced safety measures !

30.4 Advancement in skills :

Many people may be knowing , there are around 64 types of arts and advertising is considered as 65 th unofficial art ! What you are making and how you are selling totally depends on the art of advertising ! So, a producer makes a product and provide its benefits to its customer through a well explained advertisement ! Customer watches this advertisement and purchase that product ! When he uses that product , he like it and he then constantly purchase that product with or without referring to that advertisement ! So, advertisement act as first point contact between new producer and new customer ! One simple skill of making advertise consists of arranging product visuals , writing a short script , choosing the characters , filming the advertise, adding suitable music with instantly notable tag line and then broadcast in prime hours were number of people can see it !

Courtesy: CFV, Pixabay

The first level stock is made available at distributors end , once people see that advertise and decide to purchase it , demand goes high and that product clicks in the market !

So , if you see the advancement is skill of making advertisements , it is changes from paper advertisement to short clips played on national television ! This is followed by advertisements played on cable channels which is followed by those played during movie show ! In smart phone age , advertisements are available on internet , social media channels , highways and city centers ! So , who is benefitting from these skills ? Everyone ,customer , producer , advertise maker !

Likewise, many new age skills are being developed ! The person who learns these skills become the favorite of their respective recruiters !

If you consider the game of cricket , just see how many new shots are developed , how many new deliveries are developed , how many new rules are formed and how many technological inclusions are done to give decision , to provide facilities on ground and things like that !

Earlier only test was played , this is followed by formation of fifty -fifty format and now the days of twenty -twenty are going on ! The franchise model allows any player to play for any team ! Their fees are also increased and their endorsement deals are also much better that earlier ! Marketing people chosen various prominent players and actors to endorse their brands and this way of selling product become norm of selling today !

For professional developments , the short-term courses are developed with which one can add professional skills ! When you adopt those skills and implement at your work , your work completes faster and you can handle greater responsibilities ! This gives you chances of further progress and makes things easy for you !

30.5 Advancement in every field :

So , herewith we are moving towards last point of this book ! The research and development are continuous process of evolution and considering huge span of science , it's happening in every field ! The pattern of knowing a new branch of science , finding its roots to some of the existing branches , making curriculum of that branch and then awarding respective students' diploma , degree, post graduate and doctorates of that course is become modern norm of educational research !

Same thing is happening in every industry ! Creating an innovative product by using cutting edge technology , doing continuous improvements and increase its sales become a new identity of industrial research ! If a product is not giving suitable results , further modifications are done into its features and again the researched product is released to market ! This advancement is also going on in all fields ! Rarely there is any field , which is yet not accepted cutting edge technology ! And if such field is present , then it is not receiving the latest advantages of growth ! So , many organizations changed themselves according to latest happenings in the industrial research world and hence they are doing their work more easily and more professionally !

So , hope this discussion is given you fair information about research for advancement and hence constant progress !

Progress for healthy life , progress for healthy mind , progress for healthy nation , progress for healthy relations , progress for good conduct , progress for understanding inner self , progress for liberation , progress for values , progress for becoming a good research fellow , progress for becoming good research guide , progress for carrying out path braking research , progress for carrying out research for humanity ! Progress for carrying out research for good life ! Hope you have enjoyed all the steps of this book ! At the outset of its completion , sharing one simple quote of research success and that is , ' Just believe in yourself, always ! ' ✪✪✪

MULTIPLE-CHOICE QUESTIONS

1) Which of the following advancement will provide alternative manufacturing option for researchers ?

A) Advancement in material .
B) Advancement in process .
C) Advancement in Skills.
D) Advancement in services .

2) The job of scientist is to

A) Invent new things from mother nature .
B) Commercialise the inventions .
C) Understand the natural principles for human good and bad .
D) Earn international awards .

ANSWERS OF MULTIPLE-CHOICE QUESTIONS

Step 1 : Story of Curiosity !

1) A) Till the breakthrough is achieved .
2) D) All of the above
3) D) All of the above
4) C) Why earth is spherical ?
5) A) One that has saved many lives of people.

Step 2 : Story of Experiment !

1) D) All of the above
2) D) All of the above .
3) A) Aim of the experiment .
4) D) All of the above
5) A) Labelling

Step 3 : Story of Approach !

1) D) On field based
2) D) All of the above
3) D) Both A & C
4) D) All of the above !

Step 4 : Story of Will !

1) D) All of the above !
2) A) Repeat the experiment after 7th consecutive failure !

Step 5 : Story of Patience !

1) D) Taking 1000 random observations from sample size of 10 Lakhs !
2) D) All of the above !
3) A) By filling applicable system requisition on dedicated website !
4) D) A scientist with 100 patents on his name !

Step 6 : Story of Failure !

1) A) Occurrence of same defect after every trail !
2) A) Sam has performed research under Mr. Shyam , Mr. Shyam has reputation of carrying out research with most advanced laboratory equipment access in all over the world !
3) D) All of the above !

Step 7 : Story of Break & Gap !

1) B) Experiments are kept on hold for

         receiving output of other experiment for 1 day !
2) D) All of the above
3) D) All of the above
4) D) All of the above

Step 8 : Story of Logic !

1) D) All of the above !
2) A) When research logic is incorrect in the very beginning !
3) D) All of the above !
4) D) All of the above !

Step 9 : Story of Intention !

1) D) All of the above !
2) D) All of the above !

Step 10 : Story of Invention !

1) D) All of the above
2) B) Phone Pay Mobile App

Step 11 : Guide as coach

1) D) All of the above
2) D) All of the above

Step 12 : Guide as Friend

1) D) All of the above
2) D) All of the above

Step 13 : Guide as Philosopher

1) D) All of the above
2) A) For my country , product price will be according to break-even point .

Step 14 : Guide as a Question Bank

1) A) Daily Once !
2) A) Will repeat the experiment for error proofing

Step 15 : Guide as Solo Viewer

1) D) All of the above
2) D) All of the above

Step 16 : Guide as a reliever

1) D) All of the above
2) D) All of the above

Step 17 : Guide as a simplifier

1) D) All of the above
2) D) All of the above

Step 18 : Guide as a sophisticator

1) C) 5G
2) D) Satellite Surveillance

Step 19 : Guide as a presenter
1) D) All of the above
2) D) All of the above

Step 20 : Guide as a genuine citizen

1) D) All of the above
2) D) All of the above

Step 21 : Path breaking research

1) D) All of the above
2) D) All of the above

Step 22 : Impactful research

1) B) Mobile Payment Applications
2) D) COVID-19 vaccine

Step 23 : Research for Defense

1) D) All of the above
2) D) All of the above

Step 24 : Research for Business

1) A) A brand-new product market where there is no other competitor
2) A) All fields where products and services can be made available by seller – software – buyer mechanism.

Step 25 : Research for Humanity

1) D) Setting up a new venture in foreign territory , understanding detailed lawful framework , receiving various statutory and commercial permission and starting business by hiring people , buying material and machines and making customer comfortable with the new venture !

Step 26 : Research for Agriculture

1) A) Research For Soil

Step 27 : Research for Good Life

1) D) All of the above .
2) D) All of the above

Step 28 : Research for Liberation

1) D) All of the above

Step 29 : Research for Values

1) B) Business Values

Step 30 : Research for Advancement
1) A) Advancement in Material .
2) C) Understand the natural principles for human good and bad .

IT'S DONE !

www.ingramcontent.com/pod-product-compliance
Lightning Source LLC
Chambersburg PA
CBHW052137220526
45471CB00004B/1423